THE CHANGING LANDSCAPE OF PLANT SCIENCES:

Taxonomy & Ecology

A Festschrift For
Prof. S.D. Sabnis

Edited by

M. DANIEL
Reader in Botany,
Faculty of Science,
M.S. University, Baroda, 390002
INDIA

International Book Distributors
9/3, Rajpur Road, (Ist Floor)
DEHRA DUN - 248001, INDIA.

Published by:
International Book Distributors
Dehra Dun, (India)

ISBN-81-7089-223-6

Printed by: **R. P. Singh Gahlot** for
International Book Distributors, Dehra Dun- 248001 (INDIA).
and printed by Prashant Gahlot at
Valley Offset Printer and Publishers
15/2-B, Rajpur Road, Near Akash Club,
Dehra Dun - 248001.

Contents

Preface

Professor Sharad Dwarakanath Sabnis served his "Alma Mater", The M.S. University, with distinction in various capacities as a teacher, researcher, Head, Department of Botany, Dean, Faculty of Science and Pro-Vice-Chancellor. He excelled in all the posts whether they were academic or administrative. He is a visionary who was instrumental in moulding the Department of Botany to its present shape and bringing it to the national and international maps. As a scholar he excelled in the fields of floristics, chemotaxonomy, ethnobotany, ecology and environmental impact of development projects.

The publication of a Festschrift was the prime aim of all of us *i.e.*, the students, colleagues and friends of Prof. Sabnis, who met on March 15, 1994 with a view to felicitate this great scholar on his retirement. The committee, formed then, chalked out various plans which include holding a one-day national seminar besides publishing a felicitation volume. The one-day seminar and felicitation function were held on October 20th 1994, the 61st birthday of Prof. Sabnis.

The Festschrift entitled "The Changing Landscape of Plant Sciences-Taxonomy and Ecology" reflects the present scenerio in these two disciplines. Both these "classical" branches of Botany have undergone a sea of change in the past decades and emerged out as frontier disciplines in plant sciences. The age-old concepts and practices gave way to recent dynamic approaches resulting in a great resurgence of interest in these subjects. To give effective exposure to the state of the art situation, the eminent scholars in these disciplines were approached and were asked to prepare comprehensive treatises on the topics of their specialisation. The response to our invitation was overwhelming and resulted in the formulation of truly thought provoking articles. Each chapter is a combination of a review and an original article. The topics covered thus include bio-diversity, future of floristics, desert floristics, species concept, plant nomenclature, chemotaxonomy and chemical adaptations, floral anatomy, fruit anatomy, cytotaxonomy, palynology, bacterial pathogens, ethnobotany of Agustiar hills and Himalayas, endangered plants, air pollution in urban ecosystems, water relations of halophytes, inorganic fertilisers and fusarial toxins. The felicitation committee wishes to thank the contributors who complied with our demands valiantly and courteously. The valuable help rendered by Miss Julie Sebastian in compiling and proof-reading the MSS is duely acknowledged. We wish to thank our publisher, International Book Distributors, Dehra Dun for their encouragement and co-operation which led to the prompt publication of this volume.

M. Daniel

Contributors

Bhandari, M.M., Emeritus Professor, Department of Botany, J.N. Vyas University, Jodhpur - 342005, India.

Bilgrami, K.S., Emeritus Professor, Department of Botany, Bhagalpur University, Bhagalpur - 812007, India.

Bir, S.S., Emeritus Professor, Department of Botany, Punjabi University, Patiala - 147002, India.

Chaphekar, S.B., Professor and Head, Department of Botany, University of Poona, Pune - 411007, India.

Cheema, P., Department of Botany, Punjabi University, Patiala - 147002, India.

Daniel, M., Reader, Department of Botany, Faculty of Science, M.S. University, Baroda - 390002, India.

Datt, Bhaskar, Taxonomy and Herbarium Division, National Botanical Research Institute, Lucknow - 226001, India.

Dube, H.C., Professor and Head, Department of Life Sciences, Bhavnagar University, Bhavnagar - 364002, India.

Hussain, Tariq, Taxonomy and Herbarium Section, National Botanical Research Institute, Lucknow - 226001, India.

Joshi, N.C., Department of Botany, R.J. College, Bombay - 400088, India.

Karmarkar, S.M., Department of Botany, R.J. College, Bombay - 400088, India.

Kshetrapal, Shashi, Department of Botany, University of Rajasthan, Jaipur, India.

Kulkarni, A.R., Professor and Head, Department of Life Sciences, University of Bombay, Vidyanagari, Bombay - 400098, India.

Mehendra Kumar, Research Associate, Department of Botany, J.N. Vyas University, Jodhpur - 342005, India.

Mohamed, Sher, Laboratory of Plant Ecology, Department of Botany, J.N.V. University, Jodhpur - 342001, India.

Mulla, R.M., Department of Life Sciences, University of Bombay, Vidyanagari, Bombay - 400098, India.

Naik, V. N.,Professor (Retd.), Konark, 1-Kanchan Nagar, Nakshatrawadi, Aurangabad, India.

Nair, P.K.K., Director, Environmental Resources Research Centre, P.B. 1230, Peroorkada, Thiruvananthapuram - 695005, India.

Nair, Narayanan, K., Tropical Botanic Garden and Research Institute, Palode, Thiruvananthapuram - 695562, India .

Nayar, M.P., Pitamber Pant National Environment Fellow and Emeritus Scientist (TBGRI), The Greens, 19/315 Vatavila, Trivandrum - 695006, India.

Pai, R.M., Professor, Department of Botany, Dr. B.A.M. University, Aurangabad - 431004, India.

Prakash, Anand, Taxonomy and Herbarium Division, National Botanical Research Institute, Lucknow - 226002, India.

Pushpangadan, P., Director, Tropical Botanic Garden and Research Institute, Palode, Thiruvananthapuram - 695562, India.

Rajasekharan, S., Tropical Botanic Garden and Research Institute, Palode, Thiruvananthapuram - 695562, India.

Rao, R.R., Joint Director, National Botanical Research Institute Lucknow - 226001, India.

Ravi, N. Professor (Retd.), Verbena, Mundakkal Middle, Kollam, 691001, India.

Sen, David, N., Professor, P.O. Box 14, Zion, $B/B_1/A$, PWD Colony, Jodhpur - 342001, India.

Sinha, K.K., University Department of Botany, Bhagalpur University, Bhagalpur - 812007, India.

Sivarajan, V.V., Professor and Head, Department of Botany, University of Calicut, Calicut - 673635, India.

Tiagi, Y.D., Emeritus Professor, Department of Botany, M.L. Sukhadia University, Udaipur, India.

Zedah, N.H., Plant Pests and Deseases Research Institute, P.O. Box 1454, Tehran - 19395, Iran.

Chemotaxonomy in India - A Tribute to Prof. S.D. Sabnis

M. Daniel

PhytochemistryLaboratory, Department of Botany, The M.S. University of Baroda, Vadodara, 390002.

Though Baroda boasted of a sound taxonomy centre under the stewardship of Prof. Sabnis in 1960's, doing a lot of commendable work on the floristics of a number of geographic zones like Baroda, Ratanmahal, Khedbrahma, Kawant range and Savli, its forray into experimental taxonomy started by the beginning of the next decade. Establishment of a phytochemistry laboratory under an Am-PL-480 scheme enabled us to think of using the phytochemical data in classification. In the beginning we studied chromatographic profiles of some of the Cyperaceae members and found that the data obtained were useful in drawing conclusions on a number of debatable points.

The identification of the chemical markers remained elusive for quite sometime then. A number of people and Institutions helped us to isolate and characterise compounds in a microscale. The training given to the author by Dr. A. Banerji of Bio-Organic Division, BARC in 1973, can be said to be the turning point in our chemotaxonomic investigations. The Department of Chemistry of M.S. University was also very helpful to us. Dr. N.V. Mani of the same department, who collaborated with us in one research project, gave valuable contribution to standardise the methods. The help rendered by Prof. A.G.R. Nair and Prof. Sankara Subramanian of Jipmer, Pondicherry, by supplying standard samples of flavonoids also is gratefully acknowledged.

Once the methods were standardised we selected various taxa for analysis. Till today our team had analysed more than 1500 plants belonging to about 80 families and arrived at valuable conclusions. Since, explaining all the works is beyond the scope of this comprehensive review, only highlights of our investigations are presented below:

1) **Capparidaceae - Cleomaceae**

Cleome and *Gynandropsis* are more like the Brassicaceae than the Capparids, in containing flavonols such as isorhamnetin and quercetin as well as brassicine alkaloids. On the contrary, Capparids do not possess much of the flavonols and elaborated glycoflavones and proanthocyanidins instead. Therefore, the separate status of Cleomaceae closer to the Brassicaceae is upheld (Daniel and Sabnis, 1977).

2) **Malvales**

Malvales traditionally contained Malvaceae, Sterculiaceae and Tiliaceae alongwith their splinter-families Bombacaceae and Elaeocarpaceae. Except for the Elaeocarpaceae, the remaining four families possess a unique combination of characters such as stellate hairs, wedge-shaped phloem, triangular phloem rays with their broad bases towards the periphery, sepaline nectaries, valvate aestivation of corolla, cyclopropanoid fatty acids and flavones. This casts serious doubts on the placement of the Elaeocarpaceae in the Malvales. Therefore the placement of this family in the Theales is suggested to make the Malvales a very homogeneous taxon. The raising of Brownlowieae of Tiliaceae (to a separate family) and the shifting of the capsular fruited Malvacean members to Bombacaceae do not get much chemical support (Sheeja, 1991).

3) **Sapindales**

The Sapindales of Cronquist (1981), encompassing a part of the Geraniales and the Rutales of other authors, are found to be heterogeneous in their chemistry. The Rutales *sensu stricto,* consisting of the Rutaceae, Meliaceae, Cneoraceae and Simaroubaceae, are found to be highly homogeneous in producing unique nor-triterpenoids. The Burseraceae, from which the Rutales evolved, gave rise to two other groups, the Sapindales *sensu lato* and the Anacardiaceae (Umadevi, 1990). The splinter families of the Sapindaceae, the Aceraceae and Hippocastanaceae, do not appear to possess distinct identities and therefore are merged with the Sapindaceae (Umadevi *et al.,* 1986). The staphyleaceae and Melianthaceae do not share any chemical character with the above-said families and have to be placed elsewhere. The Geraniales, which arose parallel to the Rutales contain, the Geraniaceae, Oxalidaceae, Zygophyllaceae and Balsaminaceae. The chemical data support the creation of a new unigeneric family Peganaceae to incorporate *Peganum harmala* (Umadevi *et al.*, 1990). The placement of *Filicium decipiens* in the Sapindaceae (Umadevi and Daniel, 1991a), *Chloroxylon* in Rutaceae (Umadevi *et al.,* 1988), and *Balanites* in Zygophyllaceae (Umadevi and Daniel, 1991b) is upheld.

4) **Rhamnales**

The Rhamnaceae are found to be clearly distinct from the Vitaceae with which they were included (Umadevi and Daniel, 1990). Within the Vitaceae, *Leea* stands apart in producing highly hydroxylated polyphenols myricetin and gallic acid. This supports the family status given to this genus by Cronquist (1981), the Leeaceae (Umadevi and Daniel, 1991c).

5) **Fabaceae**

The difficulty in classifying the Fabaceae is evident from the large number of classificatory schemes available on them. Based on our studies, a large number of natural groups have been established. The homogeneous tribes, the circumscription of which are validated, are Abreae (with *Abrus* only, Daniel *et al.,* 1986), Vicieae, Trifolieae (Dutt *et al.,* 1986), Genisteae, Galageae (with subtribes Psoralieae, Indigoferae, Tephrosieae, Robineae and Astragalineae, Dutt *et al.,* 1987a), Hedysareae *sensu lato,* Aeschynomeneae, Stylosantheae, Desmodieae (Dutt *et al.,* 1987b), Glycineae, Erythrineae, Phaseoleae, Cajaneae (Dutt *et al.,* 1987c) and Dalbergieae (Dutt *et al.,* 1991).

6) **Gentianales**

The line of demarkation between the Apocynaceae and Asclepiadaceae being flexible due to the Periplocoideae forming a continuum in between, we picked up this group and our results prompted us to circumscribe three well-defined families Apocynaceae *sensu lato* (with only Plumieroideae), Periplocaceae (inclusive of the Echitoideae and Periplocoideae) and Asclepiadaceae *sensu stricto* (with only Cynanchoideae). The Periplocaceae formed the basic stock and the other two families diverged from them (Daniel and Sabnis, 1982). The distinctiveness of this group, in producing cardiac glycosides and typical indole alkaloids (isoprenoid), brings it closer to the Rubiaceae. Therefore, the order Apocynales is suggested with the three families Apocynaceae, Periplocaceae and Asclepiadaceae (Daniel and Sabnis, 1990). The Menyanthaceae with their alternate leaves and flavonols find an ally in the Solanaceae. *Exacum* is found to be different from the xanthone and bornesitol-rich Gentianaceae in its inability to produce these compounds and therefore a new family Exacaceae is proposed (Daniel and Sabnis, 1978). The remaining

Gentianales include the Loganiaceae, Oleaceae, Gentianaceae and Buddlejaceae. The proposed shifting of the Buddlejaceae to Scrophulariales is not favoured because the former family contains all the compounds of the Gentianaceae *sensu stricto* and does not elaborate the typical compounds of the Scrophulariales (Daniel and Sabnis, 1979).

7) Scrophulariales

Detailed studies on all the three principal families, the Acanthaceae, Scrophulariaceae and Bignoniaceae have been carried out in our laboratory. The Thunbergiaceae are found to be different from the Acanthaceae in possessing glycoflavones. The Nelsonoideae and Mendoncioideae are more or less similar to the rest of the Acanthaceae in elaborating 6-oxyflavones and iridoids, and therefore their retention within the Acanthaceae is recommended. Within the Acanthoideae, between the two taxa Imbricatae and Contorate, the former is found to be primitive due to the lesser incidence of 6-oxyflavones and iridoids. The generic and tribal limits of a number of taxa are discussed in detail (Daniel and Sabnis, 1987a; Gupte, 1994).

The Scrophulariaceae are a homogeneous taxon chemically, by the reduction/elimination of 6-deoxyflavones, flavonols and proanthocyanidins. The Orobanchaceae appear to be an offshoot from the parasitic members of the Scrophulariaceae and therefore the large number of seeds and the tannins present in this family are likely to be the secondary adaptational features. The Gesneriaceae, Lentibulariaceae and Pedaliaceae also trace their origin from the base of the Scrophulariaceae (Joseph, 1990).

The Bignoniaceae resorted to quinones in a big way especially in the tribe Tecomeae. Otherwise the family elaborate flavones. The tribes Tecomeae, Oroxyleae, Bignonieae, Jacarandeae and Cresentieae possess their own assortment of chemical characters (Padhye, 1982).

8) **Solanales**

This order contains the major families the Solanaceae, Convolvulaceae and Boraginaceae. The Salpiglossidae of the Solanaceae with their zygomorphic corolla and four stamens in didynamous condition, are considered intermediate between the Solanaceae and Scrophulariaceae and recommended to be placed in the latter family. But this group is rich in flavonols, the characteristic features of the Solanaceae and does not produce any of the advanced compounds of the Scrophulariaceae such as the iridoids and 6-Oxyflavones and therefore its retention in the Solanaceae is recommended (Joseph *et al.*, 1990).

The convolvulaceae are unique in their chemical composition producing ergot alkaloids. *Cuscuta* with its parasitic habit have evolved from the primitive members of this family. It shares the same alkaloids, but in addition contain tannins and abundant flavonoids and therefore its elevation to a separate family Cuscutaceae is favoured (Geetha *et al.*, 1986a). Within the Convolvulaceae, the generic status of *Aniseia, Merremia, Operculina* and *Xenostegia* are accepted (Geetha *et al.*, 1986b). The concept of a unigeneric tribe Evolvulae for *Evolvulus* is proposed on chemical grounds (Geetha *et al.*, 1988).

The Boraginaceae form another endline specialising in quinones. They possess flavones in addition and therefore its advanced status in the order is upheld.

9) **Lamiales**

The Verbenaceae are similar to the Bignoniaceae in opting out for quinones in a big way. Though they contain flavones, flavonols also are equally abundant here. The three tribes Verbeneae, Viticeae and Avicennieae of Briquet are chemically distinct. The elevation of Avicennieae to a separate family is not supported (Padhye *et al.*, 1990).

In contrast to the Verbenaceae, where the woody members resorted to quinones, the herbs of Lamiaceae elaborated volatile oils. This may have some relation to the geographical distribution, in that a number of members of the latter family are temperate in distribution. In flavonoid chemistry Lamiaceae possess highly methoxylated flavones and flavonols. Iridoids are almost absent in this group. (Padhye, unpublished) Studies on the fatty acids of *Ocimum* yielded valuable results useful in infrageneric classification (Dutt *et al.*, 1989).

10) **Campanulaceae**

The two subfamilies Campanuloideae and Lobelioideae are distinct from each other in that the former taxon elaborated flavonols and the latter, flavones and alkaloids. The controversial genus *Sphenoclea*, which is separated into a unigeneric family Sphenocleaceae, is similar to *Campanula* in a number of chemical characters and therefore its retention in Companulaceae is suggested (Daniel and Sabnis, 1987b)

11) **Rubiales**

The Rubiaceae, the major family within, are characterised by polyhydroxyflavonols and hydroxyquinones. Its affinity towards the Apocynales is already established. Within the Rubiaceae the two subfamilies are found to be natural groups in that the Cinchonoideae are characterised by quercetin and proanthocyanidins and the Caffeoideae by kaempferol and iridoids. The genus *Ophiorrhiza* stands apart from the rest of the Cinchonoideae in its chemistry, warranting a separate tribal status to it (Thomas, 1989).

The Caprifoliaceae are peculiar in containing both flavones and flavonols equally distributed among their members. It is distinct from the Rubiaceae in not having alkaloids, quinones and cardiac glycosides and its grouping with the Rubiaceae does not seem logical. The Caprifoliaceae find a better place in the Dipsacales than in Rubiales (Thomas *et al.,* 1988).

12) **Asteridae**

Summing up all the data obtained from the above-said families a modified scheme of classification of the Asteridae is proposed, which appeared in "Essays of Evolutionary Biology" published by Birbal Sahni Institute, Lucknow, (Daniel and Sabnis, 1992).

13) **Centrospermae**

The Centrospermae (Caryophyllales of Cronquist) are considered one of the most natural taxa in the Magnoliophyta. But within this group there exist many ambiguous groups. The studies on the phytochemicals helped us to defend the family status of Basellaceae (from Chenopodiaceae, Mangalan *et al.,* 1991), Gisekiaceae (away from the Phytolaccaceae, Aizoaceae and Molluginaceae), Illecebraceae (from Caryophyllaceae), Petiveriaceae (from Phytolacaceae) and Molluginaceae (from Aizoaceae). The concept of the two suborders Chenopodineae and Caryophyllineae is found justifiable chemically. The occurrence of DOPA, stizolobic acid, a specific enzyme involved in the biosynthesis of betalains, 4,5-extradiol deoxygenase, and p-plastids in some of the members of Fabaceae indicates a common ancestry of this family and the Caryophyllales from the Rosales. Therefore the Caryophyllidae is proposed arising from the base of the Rosideae as against from the Magnolidae, a view proposed by Cronquist in 1981 (Mangalan, 1988).

14) **Zingiberales**

The Scitaminae of Bentham & Hooker was another group studied in detail. Among the various families proposed by later workers, the Zingiberaceae, Costaceae, Cannaceae, Musaceae, Strelitziaceae, Heliconiaceae and Marantaceae were analysed for their Phytochemicals. The flavonols, alkaloids, saponins and tannins form a regular feature in these families. The family status of all these taxa are supported on chemical grounds. The Strelitziaceae appear to form the basic stock from which the various other families have evolved. The Musaceae, Heliconiaceae, Zingiberaceae and Marantaceae form independant evolutionary lines which have attained same level of advancement. The Costaceae and Cannaceae represent two further evolutionary lines emerging from the Zingiberaceae (Merh *et al.*, 1986).

15) **Cyperaceae**

Flavones were found to be the major phenolic compounds of the Cyperaceae. The two tribes Eucypereae and Scirpeae are found to be distinct, in that the former tribe contained 6-oxyflavones and the latter tricin. The proposed genera *Juncellus, Pycreus, Mariscus, Courtoisia* and *Kyllinga* are found to resemble *Cyperus* closely and therefore their merger in *Cyperus* is advocatedThe separate identities of *Bulbostylis* from *Fimbristylis* and *Fuirena* from *Scirpus* are evident in their chemistry. The family Cyperaceae is a fairly advanced taxon among the monocots (Merh *et al.*, 1987).

16) **Coniferales**

Conifers form another group which is constantly being debated in taxonomic circles. Our studies involving a number of conifers in the Himalayas revealed that the Pinaceae are clearly distinct from all others in the absence of biflavones. Instead this family elaborated stilbenes. This necessitated the separation of this family and placing it in a unifamilial order the Pinales. The Taxaceae also were found to be eligible for treating them in Taxales. The rest of the conifers were grouped in a new order Cupressales taking its name from the largest family of the group (Sheeja *et al.*, 1989).

PHYTOALEXINS

Prof. Sabnis played an active role in initiating studies on phytoalexins in the Department of Botany. Since the phyto-chemistry laboratory had developed enough expertise to deal with the secondary metabolites we thought of following up of the defensive post-infectional compounds - the phytoalexins. A number of crops have been studied for their phytoalexins elsewhere and therefore we concentrated on the trees and medicinal plants. In many of the trees it is found that the healthy plants contained methoxyflavonols/flavones and on infection with the pathogen these compounds get demethoxylated to produce highly potent antimicrobials. This phenomenon was evident in *Cassia fistula, Morinda tomentosa* (Abraham & Daniel, 1988), *Ziziphus oenopllia*, (Abraham & Daniel, 1990), *Heterophragma adenophyllum, Spathodea campanulata* etc. (Daniel, 1995). In almost all cases, the level of phenolics was much higher in the infected leaves than the healthy leaves. New phytoalexins identified included coumarins, xanthones, phenolic acids etc.

ALLELOPATHY

One of prime results of entering into the foray of Phytochemistry is the understanding of the chemical interactions between plants. The chemical mediated interactions often exert a harmful effect on a neighbouring plant. The compounds present in a plant are released to the environment by death and decomposition, leaching, root exudation and volatilisation. Every plant in an ecosystem interacts with its neighbour in a positive or negative way. In our laboratory we have studied the effects of social forestry trees which are indiscriminately introduced into any ecosystem. *Pongamia glabra* and *Cassia siamea* are found to harm the crops like rice and wheat. Other trees which are taken up for studies are *Peltophorum ferrugineum, Melia azedarach, Samanea saman* and *Delonix regia*.

Though Prof. Sabnis was instrumental in initiating all the above-mentioned disciplines, floristics remained a passion for him. The note-worthy contributions from the taxonomic group are the floras of Baroda (Sabnis, 1968), Ratan Mahal hills (Bedi, 1968), Khedbrahma (Bhatt, 1972), Savli taluka (Padate, 1974), Kawant range, (Thakkar, 1974), South-Eastern Kutch (Rao, 1981) Western Kutch (Bhatt, 1994), and Shoolpaneshwar sanctuary (Pradeep Kumar, 1994). The human pressures on the urban flora was another pet topic of Prof. Sabnis. A valuable contribution to this was done when Patil (1981) surveyed Baroda after more than 20 years of the first study (Sabnis, 1968). Another topic in which Prof. Sabnis took keen interest was the ecological impact of huge man-made projects like Narmada dam. The impact assessment of Sardar Sarovar project, which was completed in 1992, won him many laurels. Studies on ethnobotany and animal plant interactions (Chavan, 1994) also flourished at the same time.

In this short review I have made a fleeting glance of what we achieved under the dynamic leadership of Prof. Sabnis in the fields of chemotaxonomy and phytochemistry. He always encouraged people to take up new challenges for research and that is why there is an all round development in the Department of Botany. I wish him long-long years of happy and productive life.

REFERENCES

Abraham, J. and M. Daniel (1988): *Nat. Acad. Sci. Lett.* **11**:101-102.

Abraham, J. and M. Daniel (1990): *Nat. Acad. Sci. Lett.* **13**:439-441.

Bedi, S.J. (1968): *Studies on the flora of Ratan Mahal Hills*, Ph.D. Thesis, M.S. University, Baroda.

Bhatt, J. (1994): *Studies on the flora of Western Kutch.* Ph.D. Thesis, M.S. University, Baroda.

Bhatt, R.P. (1972): *Studies on the flora and vegetation of Khedbrahma region in North Gujarat with a note on the chromosome no. and karyotype of some taxa therein.* Ph.D. thesis, M.S. University, Baroda.

Chavan, S.A. (1994): *Vegetational and wild life studies in Gir forests*, Ph.D. Thesis, M.S. University, Baroda.

Cronquist, A. (1981): *An Integrated system of Classification of Flowering Plants*, Columbia University Press, New York.

Daniel, M. (1995): In *A Handbook of Phytoalexin Metabolism* and *Action* (M. Daniel and R.P. Purkayastha eds.), Marcel Dekker Inc. New. York.

Daniel, M. and S.D. Sabnis (1977): *Curr. Sci.***46**:472-474.

Daniel, M. and S.D. Sabnis (1978): *Curr. Sci.* **47**:109-111.

Daniel, M. and S.D. Sabnis (1979): *Curr. Sci.* **48**:383-385.

Daniel, M. and S.D. Sabnis (1982): *Ind. Bot. Reptr.* **I**:84-90.

Daniel, M. and S.D. Sabnis (1987a): *Proc. Ind. Acad. Sci. (Pl.Sci.)* **97**:315-323.

Daniel, M. and S.D. Sabnis (1987b): *Ind. Bot. Reptr.* **6**:53-56.

Daniel, M. and S.D. Sabnis (1990): In *Phytochemistry and Plant Taxonomy* (K.S. Bilgrami & J.V.V. Dogra eds.), CBS Publishers, Delhi, pp. 123-132.

Daniel, M., Bhatt, R.P. and S.D. Sabnis (1986): *J. Pl. Anat. Morph.* **3**:55-57.

Dutt, C.B.S., Daniel, M. and S.D. Sabnis (1986): *Biol. Bull. Ind.* **8**:181-185.

Dutt, C.B.S., Daniel, M. and S.D. Sabnis (1987a): *Nat. Acad. Sci. Lett.* **10**:83-87.

Dutt, C.B.S., Daniel, M. and S.D. Sabnis (1987b): *Acta Botanica Indica* **15**:266-269.

Dutt, C.B.S., Daniel, M. and S.D. Sabnis (1987c): *Feddes Repertorium* **98**:589-593.

Dutt, C.B.S., Daniel, M. and S.D. Sabnis (1989): *Adv. Biosci* **8**:71-75.

Dutt, C.B.S., Daniel, M. and S.D. Sabnis (1991): *In Botanical Researches of India* (N.C, Aery and B.L. Chaudhary eds.) Himanshu Publishers Udaipur, pp. 50-52.

Geetha Nair, Daniel, M. and S.D. Sabnis (1986a): *Biome* **1**:1-5.

Geetha Nair, Daniel, M. and S.D. Sabnis (1986b): *Curr. Sci.,* **55**:961-965.

Gupte, A. (1994): *Phytochemical and taxonomic studies on Acanthaceae and related taxa*, Ph.D. Thesis, M.S. University, Baroda.

Joseph, J. (1990): *Chemical and taxonomic studies in Scrophulariaceae and some allied taxa.* Ph.D. thesis, M.S. University, Baroda.

Joseph, J., Daniel, M. and S.D. Sabnis (1990): *Biol. Indica*, **I**:17-19.

Mangalan, S. (1988): *Chemosystematics of some selected taxa of the Caryophyllales*. Ph.D. Thesis, M.S. University, Baroda.

Mangalan, S., Daniel, M. and S.D. Sabnis (1991): *Biol. Indica* **2**:23-26.

Merh, P.S., Daniel, M. and S.D. Sabnis (1986): *Curr. Sci.* **55**:835-839.

Merh, P.S., Daniel, M. and S.D. Sabnis (1987): *Geobios,* **14**:217-220.

Padhye, P.M. (1982): *Chemical Systematics of Bignoniaceae and Verbenaceae*. Ph.D. Thesis, M.S. University, Baroda.

Padhye, P.M., Daniel, M. and S.D. Sabnis (1990): In *Phytochemistry and Plant Taxonomy* (K.S. Bilgrami and J.V.V. Dogra eds.) CBS Publishers, Delhi, 123-132.

Padate, S.N. (1974): *Studies on the flora and vegetation of Savli taluka*. Ph.D. Thesis, M.S. University, Baroda.

Patil, S.N. (1981): *Studies on the residual urban vegetational systems at Baroda*. Ph.D. Thesis. M.S. University, Baroda.

Pradeep Kumar (1994): *Vegetational and ecological studies of Shoolpaneshwar Wildlife sanctuary*. Ph.D. Thesis, M.S. University, Baroda.

Rao, K.S.S. (1981): *Studies on the flora of South Eastern Kutch.* Ph.D. Thesis, M.S. University, Baroda.

Sabnis, S.D. (1968): *Study of the flora and vegetation of Baroda and Environs including an account of the Cyperaceae of Gujarat.* Ph.D. Thesis, M.S. University, Baroda.

Sheeja, M.J. (1991): *Chemical and taxonomic studies on the Malvaceae and allied taxa.* Ph.D. Thesis, M.S. University,

Sheeja, M.J., Daniel, M. and S.D. Sabnis (1989): *Proc. Ind. Acad. Sci.* (Pl. Sci.) **99**:253-258.

Thomas, P.J. (1989): *Chemosystematic and pharmacognostic studies on some Indian members of the Rubiaceae*. Ph.D. Thesis, M.S. University, Baroda.

Umadevi, I. and M. Daniel (1990): *J. Phytol. Res.* **3**:1-5.

Umadevi, I. and M. Daniel (1991a): *Feddes Repertorium* **102**:607-612.

Umadevi, I. and M. Daniel (1991b): *Adv. Bio. Sci.* **10**:69-74.

Umadevi, I. and M. Daniel (1991c): *Acta Bot. Ind.* **19**:168-170.

Umadevi, I., Daniel, M. and S.D. Sabnis, (1986): *J. Pl. Anat. Morph.* **3**:169-172.

Umadevi, I., Daniel, M. and S.D. Sabnis (1988): *Feddes Repertorium* **99**:195-197.

Umadevi, I., Daniel, M. and S.D. Sabnis (1990): *Ind. J. Bot.* ***13**:23-28.*

Biodiversity and Traditional Taxonomy

V.V. Sivarajan
Department of Botany, University of Calicut - 673 635, Kerala.

Biodiversity has already become a rallying point for not only biologists, but also for other professionals, policy makers, governments, and the general public, the world over. The increasing concern over the rather rapid loss of organic diversity and the need for its conservation is amply reflected in the recently concluded Biodiversity Convention (July, 1992, Rio de Janerio). While this is, hopefully, the harbinger of a welcome change in our attitude towards nature and life, it seems doubtful whether we are adequately aware of the state of our current knowledge on biodiversity and the need for studying and understanding it better, which is an essential aspect of exploiting it for human benefit on a sustainable basis. The entire gamut of the problem cannot be discussed in detail in a short essay like this. But there are a few things which are of concern to all of us. I shall try to discuss some of them here, in three parts. The first part would deal with the importance of biodiversity and the second with the 'biodiversity crisis' and its portents for humanity. In the third part, I shall reflect on the present state of biology and would try to convince you that traditional taxonomy is the best avenue available for biodiversity studies.

BIODIVERSITY AND ITS IMPORTANCE

The World Resources Institute has defined biodiversity as "the variety of the world's organisms, including their genetic diversity and the assemblages they form. It is the blanket term for natural biological wealth that undergirds human life and its well being. The breadth of the concept reflects the interrelatedness of genes, species and eco-systems. Because genes are the components of the species and species are the components of the ecosystems, altering the make up of any level can change the others.... Species are central to the concept of biodiversity". As is obvious from this, the term 'biodiversity carries a very broad meaning, covering every component of the living system, and all forms of life, right from micro-organisms to human beings. (See Nelson & Serafin, 1992).

Discussing importance of biodiversity is a tricky proposition, because, over the centuries, we have become a highly commercialised and consumption-oriented society and cannot conceive of benefits except in monetary terms. Naturally, we have come to believe that, excepting organisms that are directly beneficial to us, rest of the biodiversity is useless. In fact, there was a time when economists believed that natural ecosystems are not very much related to the behaviour of economy and conversely, when ecologists believed that economics does not come under the purview of ecology. It was seldom realised that species are the major components of our life-supporting environment on which the survival and well being of humanity depends very much. This attitude has now been changed and we have come to realise that ecology and economics are highly interrelated. This is apparent from the recent emergence of disciplines like 'Ecological economics' (Constanza, 1991).

Assessing the value of biodiversity in economic terms is an extremely cumbersome task, because it is so pervasive and vast (for more details, see Perrings *et al.,* 1992; Ehrlich & Ehrlich, 1992). Broadly, speaking, they satisfy human needs in two different ways, direct and indirect.

The direct uses of biodiversity are of common knowledge. We all know that we depend on other organisms for almost all our daily requirements (food, clothing, shelter, medicine etc.). Also, we depend on them for planning and executing developmental programmes, say agricultural improvement, animal husbandary etc. But, the indirect uses of biodiversity often escape our attention and hence this aspect would be dealt with here in some detail. The benefits that we reap from biodiversity is the sum total of both direct and indirect uses.

It has been estimated that a single tree, over a period of about 50 years, renders goods and services worth Rupees 15.7 lakhs to the society (Das in Anil Agarwal, 1993). But, nevertheless, the indirect benefits, *i.e.* the ecological services that organisms render are of much greater value than the direct, material benefits that they provide. Mathur and Soni (in Anil Agarwal, 1993) have estimated that the material and economic benefits from a forest, over a period of 50 years would come only to about 0.6% whereas its ecological benefits may go up to a whopping 99.4 per cent. Despite this overwhelming environmental benefits, most of the current discussions on the value of bio-diversity revolve around pricing such fringe economic gains rather than the services they render. The result is that we have tended to destroy species that are not of known economic value rather indiscriminately. One of the most familiar example is that of the Earthworms which was, once upon a time, ubiquitous in our homeyards and farmlands. In fact, the ancients had realised their great potential in fertilising the land and had valued them for that. Charles Darwin was so much impressed by the formidable might of these ugly-looking, tiny creatures, that he wrote a whole book in 1881, the last of his publications. But, we were not convinced. The advent of agronomy-intensive cultivation virtually wiped them out from our fields. Now we have realised their potential and are now spending millions on research projects on them. Big companies have now taken to 'vermiculture' in a big way and the times are such that we have to purchase them for a price and re-introduce them into our fields.

The combinations of organisms (communities) and their role in bio-physical cycles in the respective ecosystems render organisms highly useful in maintaining a healthy environment. The most obvious example of such indirect benefit is that of photosynthesis, the process by which green plants trap solar energy and provide it to all other life forms in the form of chemical energy, the carbohydrates, which drive the life-processes in humans and in animals. Oxygen, which is so essential for all forms of aerobic life, is a bye-product of photosynthesis and all oxygen in the atmosphere at present has been generated through this process. Today, it is the second most common gas in our atmosphere, after nitrogen.

In the recent times, there is widespread concern over the depletion of the ozone layer. What is Ozone? It is a special type of oxygen molecule formed by three atoms of oxygen. It is formed in the stratosphere when ultraviolet solar radiations split a normal oxygen molecule and one of the resultant oxygen atoms get latched on to another oxygen molecule to form a molecule of O_3. But for this ozone layer, the earth would have been exposed to perilous solar radiations, spelling doom for all forms of life. That explains why we are so much perturbed over the situation. And mind the ozone layer would not have been there but for the oxygen that have been put in the atmosphere through photosynthetic acitivity.

Organisms those cohabit in the ecosystems and their interaction provide us with a number of such ecological services, great and small. The case of the food chain and energy cycle are well known. Besides, they help formation, binding and enrichment of soil, maintenance of water table, degradation and elimination of toxic pollutants from our environment and influence climate. Even the tiniest of organisms, has a role cut out for them in the most subtle and intricate ways. But unfortunately, we seldom realise this adequately well.

Natural systems maintain a vast genetic library which has been and will continue to be of immense value in human endeavours. Now, we have almost mastered (eventhough, there are grey areas, still) the structure and behaviour of genes and their role in determining characters in living organisms. Furthermore, we have perfected techniques to manipulate them and to produce improved new strains of beings. Superb examples of this type of manipulative technology can be seen in the area of our agricultural and medical research programmes. In fact, that is biotechnology all about. Transforming living organisms by manipulative techniques for human benefit and patenting life-forms and genes have become highly contentious issues in the recent times. It will take some time for us to understand the relative merits and demerits of this type of endeavours. But, by that time, we would have reached a point of no return, as has been exemplified by the case of the earth worms.

Apart from these, there are the aesthetic and ethical values attached to them. Other organisms have always been a part of our life. The beauty of the birds, tropical fishes, butterflies and flowering plants have encouraged and supported a variety of economic activities like bird-watching, butterfly collecting, floriculture, nature photography and films, flower export etc. They not only support economy, but also satisfy our intellectual curiosity and provide us emotional succour, warmth, solace and a healing touch. In other words, they are also objects of our heart and that is why flowers have always played a dominant role in our social mores.

The ethical values of other organisms are largely based on religious or semi-religious feelings of the various cultures. The ancient beliefs, faiths and folk-lore would testify to the fact that forests and other living beings have always played an important role in human life and activities. However, one cannot calibrate them in economic terms, because they tend to transcend to the metaphysical and spiritual aspects of man's life.

WHAT IS BIODIVERSITY CRISIS ALL ABOUT?

As humans, we are always forward-looking and on the path of progress. We call it development. We have made great strides in that direction and are now capable of turning everything to our 'benefit'. But, at what cost?. In this quest for 'progress', we have always kept ourselves on a higher pedestal, have thought of other organisms as 'out-siders' and have firmly believed that every other creation in this world is meant for man. We are on our way to the 'conquest of nature' and not on the path of compromise, living in harmony with it. We have consistently refused to believe that *Homo Sapiens* is only one of the several million species that nature has created and that nature has no special reason to be particularly benevolent to him. We have failed to take notice of the fate of the other human species, including our closest relative, *Homo africanus* which thrived on this earth once upon a time. Nature has been a creator, but it can also be a destroyer.

This egocentrism, increasing population, commercialisation and consumerism, together, have formed a highly explosive combination. The result is that we are little concerned about the natural environment and the biodiversity that it supports. Ours is the only species which is capable of tampering with and altering the living environment consciously. We have been doing that with abandon, with the help of new scientific and technological discoveries, especially in the recent past. We have mistreated nature, destroyed and degraded our forests and habitats, leading to the extinction of several species of organisms. Several thousands are threatened with extinction, not only other beings, but also several races of mankind. The Red Indians, the original inhabitants of America, now number only abou... million among a 250 million American settlers and are leading a hand-to-mouth existence in the isolated reserves in the hostile desert regions. So also, the San tribe, also called the 'Bush men', who once were the sole inhabitants of southern Africa have been similarly pushed to the Kalahari area and there are not many left. This is, probably the same case with almost all ancient tribes living in the various parts of the world, including India.

But, man has been more cruel to other forms of life which are co-existing with him here and form an integral part of his life-supporting environment, primarily by destroying forests and habitats. Thus, the estimated 7000 m. ha. of forests we had in this world in 1900 has come down to 2829 m. ha. by 1975 (cf. Khoshoo, 1986). The most recent estimate tells us that about 3 billion ha. of the land area of the world (which amounts to the area of China and India together) now stands highly degraded. Tropical forests, the centres of biodiversity and the 'cradle of evolution', have been the worst affected. They are being destroyed at the rate of 100,000 sq. kms. every year (Kothari, 1993). Situated mostly in the developing countries, almost half of such forests have already vanished and at the present pace of destruction, they may not witness the next century.

Consequently, biological extinctions, which led to the disappearance of species once in several hundred years, is now replaced by highly accelerated rates of extinctions (now almost 400 times higher than in the recent past). Thus, we have wiped out the entire species of the bird Dodo and Passenger-pegions in about a hundred years time. Today, we are loosing species at the rate of one every year. The number of species which are threatened with such untimely and unnatural extinction are far higher. 10-20 per cent of the total number of species are now considered to be endangered. If we go at this pace almost one fourth of the plant species would disappear by the second half of the 21st century. The alarming rate of such extinctions and the extent of the threat of loss of species could be roughly understood from the following conservative estimate.

Current status of species of plants and animals

Species	Total No. of sp.	Extinct	Endangered
Plants	2,50,000	384	18,964
Fish	19,056	23	320
Amphibians	4,184	2	48
Reptiles	6,300	21	1,355
Birds	9,198	113	924
Mammals	4,170	83	414

Note: 'Extinct' means species which have disappeared since 1600 A.D. (from McNeely *et al.*, 1990).

According to the most recent estimate by the World Conservation Monitoring Centre, Cambridge, 41,000 plant species and 6,000 animal species are now facing the threat of extinction.

India is one of the twelve megacentres of biodiversity and has two top priority 'biodiversity hotspots' with high degree of endemism, within its boundaries. It has a unique biogeographic composition, comprising of Palaearctic, Indomalayan and African elements. With just 2 per cent of the total land area, it supports 5.2 percent of the total plant species and 4.3 per cent of the world's animal taxa. The total number of plant species in India would come to about 45,000 including 15,000 species of Flowering Plants, 600 ferns, 2100 Bryophytes, 5000 algae, 20,000 fungi and 1500 species of lichens. Among animals, there are 340 species of mammals, 1200 birds, 420 reptiles and amphibians, 2000 fishes, 4000 molluscs and 50,000 species of insects. India has 75 per cent of the 425 families of Flowering Plants in the world (Punita, 1993).

But, our track record in forest conservation, the store-house of biodiversity, has been rather dismal. In India, the area under forest cover has rapidly dwindled to a mere 17 per cent. As a result, several hundred Indian species are now on the verge of extinction and 3000-4000 species of flowering plants are now under threat. Among animals, the coconutcrab of Andamans, Lion-tailed monkeys, striped goats and the Malabar large spotted civets of Western Ghats and the musk deer, mouse deer and the snow leopard of the Himalayas are facing extinction. Out of the 19 species of the primates, 12 are now endangered.

Forests have been our hunting grounds for medicinal plants, traditionally. But, several of our medicinal plants have disappeared or have become extremely rare due to forest destruction. Thus, *Rauvolfia serpentina, Cöptis teeta, Commiphora mukul, Rubia cordifolia, Nardostachys jatamansi* etc., which are of great medicinal value are now in peril. In Kerala alone where we have about 540 medicinal plant species, many are now extremely rare and they include *Coscinium fenestratum, Mahonia leschenaultii, Kaempferia galanga* etc. (See, Anonymous, 1991).

This planet is also being improverished by the loss of races and varieties of domesticated plants and their wild relatives, which would have serious repercussions on our future agricultural and medical research and development. Out of the 80,000 species of edible plants, at present we are using only 30 species as major sources of food and 90 per cent of our food requirements are met by three species, wheat, rice and maize. Many of them could have come to our help in future, but we don't care. On the other hand, the 'green revolution' has ushered us into a era of genetically uniform, improved strains of crop plants. This has resulted in the loss of the valuable traditional varieties which have stood the test of time. Thus, very few of the 47,000 traditional varieties of rice is cultivated at present. The wild relatives of crops - cereals, millets, legumes, oil plants - dwelling in our forests have also disappeared very rapidly along with the forests. We have failed to learn our lessons from the past. It is worth remembering that when millions in Europe perished due to the infamous potato famine, it was a wild species from the Amazon basin, *Solanum berthiautii*, which saved the crop. So also, when the paddy crop in South East Asia was devasted by a stem-borer in the 1970's, our researchers ultimately found the resistant genes in some of the traditional varieties cultivated in Kerala.

Similarly, plant medicine has evoked great interest among people throughout the world, recently. Plants have always helped us in alleviating human sufferings and diseases. But so far, only 5-10 per cent of our higher plants have been investigated for their medicinal prowess. They still remain the "sleeping giant of drug development" (Farnsworth & Morris, 1976). At a time when cures for diseases like cancer are elusive and new diseases like AIDS are rearing their head, spelling doom for humanity, these plant resources could be immensely useful (See also Akerele *et al.*, 1991).

So, the point is that, as we progress, we are going to be more dependent upon the biodiversity and that is why Prof. Raven (1985) said that the key for future quality of human life on this planet lies with the existing biodiversity. But the million dollar question is: whether we would go on with the predation of nature or should we put a full stop, or at least a semicolon to it. It is a choice between the survival and suicide of the humanity.

The most important stumbling block on the way to biodiversity conservation is the mode of our developmental programmes themselves. We destroy our forests and habitats, overexploit our biological resources, all for our development. But, how long can the nature survive this onslaught? The world has now realised the perils of such progress and are now pleading the case of what is called 'sustainable development'. In short, this calls for ecological prudence. The most important point here, for all of us to remember, is that we cannot thrive here by eating up the capital, instead we have to be contented by feeding on the interest (Gadgil, 1985).

For 'sustainable development' it is not enough that we conserve biodiversity. It is equally important to acquire a reasonably good knowledge about it so that we can use them for human benefit in various ways, albeit prudently. Since there are different levels in biodiversity - morphological, physiological, behavioural, biochemical, genetic etc. - we have to decide upon the priorities first, if we have to get a real and complete picture of the extent of biodiversity that is existing at present. In this context, nobody would doubt that the most basic requirement is to understand the diversity at the morphological level, because that is what helps us in recognising natural groups and organisms as wholesome beings in the first instance. Traditional taxonomy is the subject that deals with this aspect and so, let us take a brief look at it.

TRADITIONAL TAXONOMY: THE BEST AVENUE FOR UNDERSTANDING BIODIVERSITY

Most biologists, other than taxonomists, consider as "tax on me" and as a subject which has crossed its 'expiry date' long back. Some of them even refuse to consider it to be a science at all, but an 'esoteric art' (Kubitzki, 1977) and criticise it for various faults, some of them real and some imaginary. They are not convinced that taxonomy has any worthwhile role to play in the progress of biology or in solving human problems, except for identifying and naming plants, which they think, is irrelevant. This blasphemy and irreverance to taxonomy has its roots in the present mode of imparting taxonomy teaching so much so that somebody has remarked that it is a subject that is 'badly learnt and worse taught'. It is time for us, taxonomists, to take stock of the situation, to have an introspection as to why people consider this prestine science to be a lame duck and to seek ways and means of redeeming it from this callous disregard. Problems that the teachers and students of this subject, the worldover, are facing have been dealt with in some detail in an earlier paper (Sivarajan, 1992). Here, I would like to dwell upon some other broader issues that turn biologists away from the perimeters of taxonomy.

If we take a close look at it, there is no need for us to be offended by the equation of taxonomy with art. In fact, there are certain things in common for them: both are strongly aesthetic. Aesthetic appreciation chiefly rests on the sense of form, the co-ordination of various elements into a whole which is mentally satisfying and harmonious, giving the impression of completeness. This sense of form is very much a part of classification (in so far as its primary objective is recognition of natural taxa) as well as artistic endeavours, and hence they are related. Both taxonomists and artists are seeking patterns within diversity, but taxonomists record the patterns which (they think) they see in nature while artists record those which they find in their own imagination (Stearn, 1964).

But, despite this aesthetic ties between art and taxonomy, the latter is, unmistakably, a science. It has always been progressive as other sciences are. Arts are less so. An artist often flies backward because he does not care where he is going, only about where he has been (Frye, 1981). Nevertheless, taxonomy is not an experimental science as physics, chemistry or molecular biology and so one cannot repeat, in the same sense, the construction of classifications and test the predictions derived from them (as demanded by the so-called 'scientific method'). But, it is a science in the sense that it endeavours to provide systematised information on natural groups of organisms, to develop hypotheses and to devise tests for their disproof or invalidation (Stuessy, 1990).

Taxonomy, has its own problems, no doubt. We are not sure about the theoretical frame work that would be required for a stable and useful classification, nor do all of us agree on a uniform methodology so much so that Heywood (1974) compared systematics with the 'Stone of Sisyphus'. It is a house highly divided on various lines. Consequently, instead of a wholesome taxonomy, we have, now several different taxonomies like artificial, natural, evolutionary and phylogenetic demarkated on the theoretical foundations and classical taxonomy and biosystematics based on methodological differences. In biosystematics, again, we have different subdisciplines like cytotaxonomy, chemotaxonomy, macromolecular systematics and so on. Even now, we are groping in the dark as far as definitions of taxonomic categories are concerened. We don't even have a universally accepted concept of species, the basic unit in taxonomy. Then, there are the ubiquitious 'splitters' and 'lumpers' who wreck havoc with our concepts of natural groups and their delimitation. Finally, even as the botanical names themselves are cumbersome for other biologists, the frequent changes in names of taxa add fuel to the fire (for a detailed discussion on these aspects, see Sivarajan & Manilal 1981, Sivarajan 1989, 1991, 1992 and Manilal & Sivarajan, 1991) and turn young aspirants away from this area in search of greener pastures. As a result, the number of taxonomists have been steadily declining to the extent that there are not more than 4000 'good' taxonomists in the world now, of which only 1500 are taxonomists competent to deal with tropical organisms (See Wilson, 1988). The irony is that 40-50% of the total number of species or organisms are sheltered in the tropical belt and yet most of the tropical taxonomists hail not from this area but are living in the advanced temperate countries of Europe and America. This would explain why an inventory of tropical biodiversity has, so far, remained an elusive goal (See Prance, 1977, 1984). The number of taxonomists given above might be an underestimate smacking of an european bias, yet it is true that most of our budding biologists do not, usually, opt for a career in taxonomy. But WHY?

First of all, illogically though, we have traditionally recognised two different types of sciences: pure or basic and applied. The goal of pure science, of which taxonomy is considered one, is to further our understanding of the universe. This sort of scientific pursuits may not have any immediate applied value. This is the type of work that most of us, in the universities and herbaria are doing. As a result, most of us do not have anything to show to the world, except our publications in which nobody except other taxonomists are normally interested. Naturally taxonomists, remain unsung and uncelebrated.

Applied sciences, on the other hand, address themselves to everyday practical problems of man and have commercial objectives than philosophical. Consequently, people generally think that basic scientists are ivory-tower dwellers having no concern for their burning problems and whatever they do are irrelevant to them. But, they readily accept applied scientists and esteem them for what they do. This dichotomy between pure and applied science is deeply entrenched in our psyche.

But, a discerning mind would easily find that this antagonism is unwarranted. Most applied sciences and technological advances have their roots in basic sciences, which are themselves fountainheads of discoveries. Isaac Newton who discovered gravity and Einstein who propounded the theory of relativity were basic scientists, yet their contributions have revolutionised the science and the whole world. So also, the basic scientific endeavours in the areas of cytology, genetics etc. have been the harbingers of the much acclaimed 'green revolution' of which we all have been beneficiaries. So, just as we don't have anything valuable and valueless in science, we also don't have pure and applied sciences in the strict sense of these terms. Basic science can have applied values, but the only thing is that we will not be able to forsee them.

Taxonomy, the earliest of all biological disciplines, is such a science. To a casual observer it may be a pure science, but it can have immense applied value as well. Unfortunately, most people consider plant hunting and the entire gamut of classification to be an academic excercise useful only to satiate the intellectual curiosity of the taxonomists, which it is not. That would be obvious if we take a look at the development of botany during the colonial periods (18th and 19th centuries) when plant explorations and taxonomy involved a lot of commercial considerations such as spices trade, introduction of plantation and other agricultural crops and the demand for horticulturally interesting plants. In fact, the support for botanical institutions like Kew, Leiden and Paris for plant explorations stemmed largely from the immense benefits that the respective governments and the people reaped from such endeavours in agricultural and other economic enterprises, which *inter alia* permitted some scientific and academic pursuits also on its way (Heywood, 1988; Stearn, 1988).

In hindsight, now we are realising that taxonomy, that is the study of biodiversity, can be of great value not only for the progress of biology, but also for mitigating human problems (See also Hedberg, 1988). In fact, taxonomy is a two-in-one. It is a basic science to the extent that it enables us to understand the natural diversity and its evolution better, but it is also an applied science as it provides a basic framework from which other applied sciences can draw heavily.

Man has always been baffled by mysteries of life. Earlier, we were given to understand that the mystery lies in the diversity of life itself. But, later we discovered that life is basically uniform. Consequently, we have given up the traditional wisdom and have been lulled into thinking that the mysteries do not lie in the diversity in appearance, but in the biomolecules that underline the basic uniformity. We have made great leaps on this front of understanding organisms through their molecules, no doubt. We know that every cell of a house mouse has four strings of DNA, each with a billion nucleotides organised into a hundred thousand structural genes, that if fully stretched they will be 1 m long and if magnified to the size of an ordinary wrapping string they will be around 950 kms long. We even know that the genetic information in this cell, if translated into ordinary sized printing letters, will fill all the fifteen editions of Encyclopaedia Britannica published since 1768 (Wilson, 1988). The progress of biology has, indeed, been phenomenal.

But the question is: Are we anywhere near solving the mystery of life? In spite of great developments, what do we know about organisms as wholesome beings and their existing diversity? One may have to concede that our record on this track has not been all that good. We even do not know how many species of organisms are there on this earth co-existing with us. At best, we would be able to quote that it is somewhere between three and ten million, but then the difference of 7 million indicates our area of ignorance in this respect. Let us accept it for argument sake. Then, have all of them been discovered and described? No. So far we have discovered and described only ten to fifteen per cent of them. And we have a reasonably good knowledge about only one per cent of the described ones. Even as this points to the lacunae in the present day biology and its lop-sided development, the most recent assertion by an entomologist, Erwin (1983), would take the wind out of the sail of those who firmly believe in this estimate. Erwin, after protracted studies of tropical insects, has come to the conclusion that there are at least thirty million species of insects alone, most of them living under tropical forest canopies. So, where do we stand in our understanding of biodiversity? Yet, the people, by and large, are given the impression that the biodiversity is already catalogued. Of course, the number of species described is quite impressive - about 1.4 million since Linnaeus - and might appear to be complete. The corollary would be that systematics, defined broadly as the description and analysis of biodiversity, is an old-fashioned science concerned only with routine tasks (Wilson, 1988). But considering the scenario outlined above, this is naivity undiluted. The large number of species that are yet to be discovered and described, discerning their natural relationships and classifying them ensures taxonomy a continuing role in biology. This task and hence taxonomic pursuits are all the more urgent and important at present, because food, medicine and other human requirements would be in short supply in the coming decades and hence there is a great need for augmenting the resource base, *i.e.*, plants.

Taking about life and its mysteries, one should keep in mind that there has always been a fundamental difference between the oriental and occidental approach to life. The easterners, especially we Indians, have traditionally visualised organisms as highly intergrated, functional, wholesome beings while the western people have a mechanistic approach which looks at them only from their structural entirety. One can easily make it out if we take a look at the oriental and western medicine. While the former looks at the whole individual as a complete 'ecosystem', the latter, as would be evident from the innumerable specialities and superspecialities that we now have, visualise him as an assembly of its structural parts. This has a lot to do with the way in which biology has 'progressed'. Modern biology, again, has specialities like morphology, anatomy, physiology, genetics, biochemistry and what not. In biology, we have been studying organisms in bits and pieces and the wholesome approach has been lost sight of.

Eversince we have discovered the basic unity in life at the genetic level, we have been under the impression that understanding genes is the panacea for all ills of biology. Such studies have helped us, indeed, in understanding life better, but it had its shortcomings as well which we did not realise. This has recently been demonstrated by some studies in human beings which suggest that genes are not the only factors controlling character transmission. It was observed that the same mutation on chromosome no. 15 of human beings would result in two different syndromes - Angelman syndrome and Prader-Willi Syndrome - in the offsprings, depending upon whether they inherit it from the father or the mother.

But the point more relevant to our discussion here is that these protagonists of the unity (rather than diversity) in life think that because plant and animal life are basically the same, zoology and botany can well be combined into a broad biology. That is why, we now have schools of life-sciences or biosciences in many of our universities, instead of departments of schools of botany and zoology. Looking at it, it would appear that if these subjects could be limited to the study of common denominators or fundamental unity of life in plants and animals, we could, indeed have been spared of the burden of so much of publications on each; but this could have resulted in terrible loss of vital information and in their special appeal to human mind. Potato and Tomato may be very similar plants for a botanist, but, the importance of each rests in the features that differentiate them (Crownson, 1970). In short, the study of diversity of life is, at least, an equally important aspect of the science of life as is the study of their basic unity.

So, how can we appreciate the diversity of organisms in the best possible way? The question is especially relevant now when we have different types of taxonomies. Broadly, we have two different approaches: the classical approach which emphasises the external appearance, the phenotype, of the organisms and the biosystematic or experimental approach which rely on experimental characters including molecular ones, often without bothering to correlate them with phenotypic charcters. Such studies have been useful in limited ways, especially to test the various hypothesis inherent in the classificatory systems (Gilmartin, 1986), but have not been as useful in constructing classifications. Moreover many of these characters which were once thought to be infallible and of better taxonomic value, have now been found to be at least as variable as morphological features. It has also been found that convergent evolution is almost as rampant here as is in morphology. In short, biosystematics have created more problems than it has solved (See Symposium on "Biosystematics at cross roads' in *Taxon 19*:137-214. 1970: Hagen, 1983).

In fact, it is becoming increasingly clear that with these genetic molecular approach, we cannot trivialise morphology in taxonomy, primarily because we are classifying organisms and they are recognised by their morphology (Heywood, 1977). Genes are important, no doubt, but their importance lies in what they do to the organisms (Cronquist, 1975); and so, unless we carefully correlate genetic or biosystematic characters with morphological attributes, we would not be able to appreciate the entire gamut of organic diversity and cannot construct meaningful and useful classification of organisms. Cronquist (1975) had already warned us about the pitfalls of overly relying on genetic charcters for taxonomic purposes. So, taxonomy seems to be heading back to the traditional, morphological taxonomy.

It seems to me that nobody would doubt that the most intelligent way of understanding, analysing and cataloguing biodiversity at the basic and existing level, is by looking at their general appearance. In fact since very early times, man has recognised natural groups that way. It is the same way, taxonomy also developed until the 1900s, when the goal of taxonomy was rewritten (Huxley, 1940). This led to the exodus of taxonomists from their familiar home ground to the rather unfamiliar and rugged terrain of evolution, leaving classical taxonomy poorer. Now, the tables are being turned. Languishing in ignomity and disregard all these times, traditional taxonomists, now find that their subject is gaining ground. The demand for trained taxonomists is increasing as drug prospectors, ethnographers, plant and animal collectors, geneticists and breeders are exploring the biotic resources of the globe, especially the tropical parts.

To my mind, the traditional taxonomy, so far as its objective is to recognise natural groups, is the best bet for macrolevel biodiversity studies. It is not my intention to belittle; the contribution of biosystematics to taxomomy, but certainly, I am of the opinion that the centre-piece shoud be morphology, because natural groups are not recognised by their genes or molecules, but their external form and appearance. Classical taxonomy is not yet a lame duck, as many consider it to be instead, taxonomy is not yet a lame duck, as many consider it to be. One only wishes that there are enough people to take care of it. I conclude this with a quote from the famous Cambridge Botanist, S.M Walters (1989).

"There are always those who find the game too pedestrian and for such people, a science nearer to what they perceive as frontiers of knowledge will be more attractive. But, I take heart from the fact that there are always, in each generation, botanists fascinated by variety and diversity, who like the challenge of ordering of nature". It is these people who would bear the torch of biodiversity and taxonomy for posterity.

REFERENCES

Akerele, O., Heywood, V.H. & H. Synge (1991): *Conservation of Medicinal Plants,* Cambridge.

Anil Agarwal (Ed.) (1993): *The Price of Forests*, New Delhi.

Anonymous (1991): *Proceedings of the Symposium on rare, endangered and endemic plants of the Western Ghats,* Thiruvananthapuram.

Constanza, R. (Ed.) (1991): *Ecological Economics: The science and Management of sustainability,* New York.

Cronquist, A. (1975): *Ann. Miss. Bot. Gard.* **62**:517-520.

Crowson, R.A. (1970): *Classification and Biology*, London.

Ehrlich, P.R. & A.H. Ehrlich (1992): *Ambio* 21(3):219-226.

Erwin, T.L. (1983): In, Sutton, L., Whitemore, T.C. and Chadwick, A.C. (Eds.) *Tropical Rain Forest: Ecology and Management*, Edinburgh. pp. 59-75.

Farnsworth, N.R. & R.W. Morris (1976): *Amer. J. Pharm.* **147**:42-52.

Frye, N. (1981): *Science* **212**:127-132.

Gadgil, M. (1985): *Landscape Planning* **12**:285-299.

Gilmartin, A.J. (1986): *Taxon* **35**:111. 8-119.

Hagen, J.B. (1983): *Taxon* **32**:406-416.

Hedberg, (1988): *Symb. Bot. Ups.* **28**(3):1-3.

Heywood, V.H. (1974): *Biol. J. Linn. Soc.* **6**:169-178.

Heywood, V.H. (1977): *Pl. Syst. Evol. Suppl.* **1**:1-12.

Heywood, V.H. (1988): *Symb. Bot. Ups.*, **28**(3):21-28.

Huxley, J. (1940): *The New Systematics*, London.

Khoshoo, T.N. (1986): *Environmental priorities in India and sustainable development*, New Delhi.

Kothari Ashish (1993): *The Hindu Surv. Environ.* 1993:44-48.

Kubitzki, K. (1977): *Pl. Syst. Evol. Suppl.* 1:21-31.

McNeely, J.A., Miller, K.R.; Reid, W.B.; Mittermeier, R.A. & T.B. Werner (1990): *Conserving the World's Biodiversity*, Washington DC.

Manilal, K.S. and V.V. Sivarajan (1991): In, Aery, A.C. & Chaudhary, B.L. (Eds.) *Botanical Researches in India*, Udaipur, pp. 3-8.

Nelson, J.G. & R. Serafin (1992): *Ambio* **21**(3).

Perrings, C., Folke, C. & Karl-Goran Maler (1992): *Ambio* **21**(3):201-211.

Prance, G.T. (1977): *Ann. Miss. Bot. Gard.* **64**:659-684.

Prance, G.T. (1984): In, V.H. Heywood & D.M. Moore (Eds.) *Current Concepts in Plant Taxonomy* London. pp. 365-396.

Punita, J.C. (1993): *The Hindu Surv. Environ* 1993:49-51.

Raven, P.H. (1985): *Boil. Edn.* **5**(3):5-7.

Sivarajan, V.V. (1988): In "*Role of Herbaria in Scientific research and development*" Rajasthan Univ., Jaipur.

Sivarajan, V.V. (1991): *Introduction to the principles of Plant taxonomy*, Cambridge & Delhi (2nd edition).

Sivarajan, V.V. (1992): In *"Problems and prospects of Angiosperm taxonomy in India*, Jiwali Univ. Gwalior.

Sivarajan, V.V. and K.S. Manilal (1981): *Recent Trends in Life Sciences* **9**:75-85.

Stearn, W.T. (1964): In Heywood, V.H. & McNeill, J. (Eds.) *Phenetic* and *Phylogenetic classification,* London. pp. 83-86.

Stearn, W.T. (1988): *Symb. Bot. Ups.* **28**(3):40-47.

Stuessy, T.F. (1990): *Plant Taxonomy*, New York.

Walters, S.M. (1989): *Pl. Syst. Evol.* **167**:35-41.

Wilson, E.O. (1988): *Symb. Bot. Ups.* **28**(3):5-12.

Future of Floristics in India

V.N. Naik

"Konark", 1-Kanchan Nagar, Nakshatrawadi, Aurangabad.

Floristic studies, as defined by Lawrence (1951), are *taxonomic studies of a flora, or of a major segment of a flora, of a given area*. The studies may range in extent from a compiled check list of vascular plants of a small politically bounded area to a thorough taxonomic or biosystematic analysis of the components of the vascular flora of a continent. Field studies and personal observation of plants is the basic approach to a thorough inventory of the plants involved. The great Rishies (meaning the researchers) of ancient India, particularly of the Vedic period (approx. 2000 to 5000 year B.C.) devoted their entire lives in the careful and contempletory studies of plants. Their main interest was to know plants in totality along with their potential economic importance. But unfortunately, these achievements in science have not found fitting recognition in the history of science. This has happened, according to Sivarajan (1984) and Sivarajan and Robson (1991), for two reasons. Firstly due to the lack of knowledge of the Sanskrit language amongst the westerners and secondly, no serious attempts were made by the Indians to highlight this ancient Indian knowledge. A few names may be mentioned who attempted to bring to light many of the ancient Indian ideas and concepts of biology. These include Seal (1915), Howard (1926), Majumdar (1927), Chaudhary (1932), Sircar (1950), Hora (1953), Satya Prakash (1965) and Subramanyam (1979).

The available Sanskrit literature of ancient India on plant science indicates that during the Vedic period, the Indians had a sound knowledge of *Vrikshayurveda*, the science of plant life. The detailed account of the ancient Indian literature and achievements in plant science can be traced in *Parashar's Vrikshayurveda*, writings of

Uddalaka and *Yadnavalka*, *Varahmihira's Brihatsamhita* and *Agnipurana, Sharangdhara's Vrikshayurveda, Charaka Samhita* and *Atreya samhita*. These early Hindu treatises deal with plants mainly in relation to medicine, agriculture and horticulture. It is said that some 2500 years ago, Bhikshu Atreya, a professor of the University of Takshasheela, asked his pupil Jivaka to collect, identify and describe the properties of all the plants, growing within a distance of four *yojanas* from the university. (Subramanyam, 1979). Dhanvantari, Nagarjun, Sushrut are some more names of ancient botanists who had an intimate knowledge of medicinal plants. The famous medicinal plant *Rauwolfia serpentina* (L.) Benth. ex Kurz (Apocynaceae) was described by Charaka under the Sanskrit name *Sarpagandha*, which is useful as an antidote for snake-bite and insect stings. It also was known to have curative effects in cases of insanity.

A well developed terminology to describe different kinds of plants and plant parts together with various phases of plant development was also available. The ancient Indians had a good knowledge of *ankurbheda* (seed germination) and conditions necessary for it. They could differentiate between *aprakanda* (caulescent) and *stambha* (acaulescent) plants, *Skandha shakha* (primary branches) and *prashakha* (secondary branches), *vrikshadani* (parasites) and *vriksharuha* (epiphytes) etc.

They recognised *dwipatra* (bifid leaves of *Bauhinia*), *triptra* (3-foliolate leaves of *Aegle marmelos*) as well as a whorl of seven leaves, *Saptaparna*, as in *Alstonia scholaris*. Different shapes of leaves, forms of inflorescence, flower and forms of various floral organs such as *pushpadala* (petals), keshar (stamens) and *renu* or *paraga* (pollen grains), different types of fruits and variety of seeds were employed in the classification of plants. They also had a knowledge of various kinds of propagation, e.g. *Bija ruha* (seeds), *mulaja* (root's), *skandhaja* (stem), *parna* bija leaf), *agrabija* (budding) etc.

The systematic account of certain well defined groups may be found in *Vrikshayurveda* together with the knowledge of soil types, plant diseases, plant anatomy, the role of leaves and their relation with the sunlight etc. This treatise by *Parashara*, was the basis of botanical teaching preparatory to medical studies in those days. This treatise also deals with the external morphology, description of fourteen forest types in India and the internal structure of leaves, cells etc. A system of classification based on a comparative morphology of plants was provided. A large number of families (*ganas*) were also clearly defined as to be easily recognisable today (Majumdar, 1927, 1946). The plant classification was based on three major considerations namely, i) *Udvidha* (botanical), ii) *Virechanadi* (medicinal) and iii) *Annapanadi* (dietic). The botanical classification by *Manu*, the famous author of *Manusmriti*, recognised *Oushadhis* or annuals, *Vanaspatis* or cryptogams, *Vrikshas* or trees, *guchha* (bushy herbs), *gulmas* (succulents), *trinas* (grasses), *pratanas* (creepers), *Vellis* (climbers) etc. Thus, the classification of this type was based on habit.

The highly effective and very ancient system of medicine, the *Ayurveda*, is based on medicinal properties of plants. *Agnivesha, Jatukarna, Bhela, Harita* together with *Charaka, Dhanwantari, Sushruta, Nagarajuna* etc. are the notable names in this regard. These scientists used to hold symposia on medicinal plants and their uses. The first such symposium was held somewhere in the Himalayan region during the 7th century B.C. and was presided over by the great sage *Bharadwaj*. The account of this symposium is given in the *Charaka Samhita* (c.f. Satya Prakash, 1965).

In one of the classifications Charaka recognised the following six *vargas* (groups) based on their dietic value:

i) *Saka - dhanya varga* (Cereal grains)

ii) *Sami - dhanya varga* (pulses)

iii) *Shaka varga* (leafy vegetables)
iv) *Phala varga* (fruits)
v) *Harit - varga* (green vegetables) and
vi) *Adharayogi varga* (oil seeds).

Sushruta's classification is even more elaborate in which he recognised 15 groups.

It is therefore amply clear that the ancient Indians had made substantial progress in plant sciences (Majumdar 1927). But it is rather disheartening to read from Burkill (1965) that the plant science was foreign to India.

It is a fact that for a long period between seventh and sixteenth century A.D. there has been no progress in plant science worth noting. The interest in this discipline was revived on the arrival of foreigners on the scene. This progress of plant exploration is well documented by Burkill in the form of history of Botany (1965), Subramanyam (1979), Stewart (1979, 1982) and Basak (1981, 1983).

The early explorations during modern times were made by the Portugese and Dutch. Garcia d'Orta was probably the pioneer amongst the European plant explorers who gathered information on Indian medicinal plants and published in his *Os Coloquis* in 1563. This can be said to be the first printed publication on Indian plants. This was followed by the publication of 12 volume work, *Hortus Malabaricus* by Heinrich Van Rheede in Amsterdam (1678-1693). The great advance was made later by the Tranquebar missionaries and other Europeans. It must be pointed out here that these plant explorers included a few botanists and many others like physicians, surgeons, army officers, foresters, mathematicians and even the administrators. The collections made by most of these explorers were transported to their mother lands while only a few specimens were housed in the herbarium at Calcutta.

The first voluminous work published on the plants of India is the 7-volume "*Flora of British India*" by J.D. Hooker (1872-1897). This was followed by several regional floras also written and published by the Europeans. There is no doubt that these foreigners contributed significantly to our knowledge of Indian plants. In fact, their publications will continue to be the reference works as long as this activity of plant exploration will go on in this country.

Although most of these publications are commendable and of great value, it must be remembered that most of the authors did not have personal experience with the living plants. Authors of the floras of this period drew their information either from the herbarium material or from the specimens collected by their collectors. Thus, these floras lack first hand field observations. Even the descriptions and the terminology used in some of the works are far from satisfactory. Theodore Cooke, for instance, was an engineer looking after the Pune - Bombay railway tract. His "*Flora of the Bombay Presidency*" still continues as a reference work for any floristic investigation in the region. But it should be noted that the descriptions given by Cooke are not original, the terminology is not botanically accurate and there are quite a few misidentifications. It is due to this reason that many later explorers like Blatter, McCann, Santapau and others had to revise this flora, more or less thoroughly. Many other regional floras and even the voluminous Hooker's flora are in need of a drastic revision. All India symposium on Floristic studies in India, held during November 1977 was attended by a large number of workers in this field. Almost all of them expressed a need for thorough exploration of various regions in our country (c.f. *Bull. Bot. Surv. India* vol. 19, Nos. 1-4 published in 1979). Since then, however, many publications have appeared on the floras of underexplored regions. Santapau's *Flora* of *Khandala* (1953), *Flora* of *Purandhar* (1957), Maheshwari's *Flora* of *Delhi* (1963), Bhandari's *Flora* of *the Indian Desert* (1978), Saldanha & Nicolson's *Flora of Hassan District* (1976), Manilal and

Sivarajan's *Flora of Calicut* (1982), Mathew's *Flora of Tamilnadu Carnatic* (1983) are some of the excellent contributions by the Indian researchers in the post independence period.

This progress, however, did not proceed well in the later period. Although there are a few notable exceptions, the field botany, in general, suffered a serious set back under the deceptive cover of modern disciplines in botany. More and more botanists developed a feeling that this plant science of collection, naming and classification is getting out-dated and the disciplines such as genetic engineering, molecular biology, biochemistry, environmental science, biotechnology and such others are gaining grounds.

A few sincere investigators did find a head-way in their disciplines but soon it was realised that these modern disciplines are limited by sophisticated equipments, costly apparatus and relatively poor financial support. The present status of these "modern" disciplines in this country is greately disappointing. Thus, we have reached a stage where we are unable to proceed with the modern disciplines and at the same time do not like to go back to the classical or basic disciplines. As a result, the botanical studies in general are facing a stalemate. Most of the researchers, investigators and the young scientists are not in a position to pursue their investigations in a desirable way. In order to survive and maintain their social status, wilingly or unwillingly, they are adopting manipulative short cuts and are trying to deceive others. Many of them have now taken up investigations in the field of floristics but without any experience or training in this line. The Botanical survey provided facilities and monitory help for University and college teachers to undertake district flora work. Many of these teachers never had a basic training or even liking for such a type of work. Naturally, they themselves have never gone to the field and their students also were not serious about collection of plant material. The floras published by the Botanical

Survey in the recent years include a large number of such investigations. They are neither the results of honest work nor are exhaustive. Many of the identifications are doubtful and there is no way of checking them as the collections are not well-preserved. Thus, these floras, apart from bringing some credit to authors, do not add to any positive contribution. They are mostly unreliable and/or far from complete.

In the teaching institutes, it is now very difficult to find a teacher in botany who can identify a plant correctly. Most of the Universities and colleges neither have plant taxonomy as a compulsory subject nor have personnel to teach the subject. In a few exceptional cases, plant taxonomy courses are being conducted single handedly. The class-work material and the preserved plant material in many institutions are incorrectly identified and there is no effort or willingness to improve the situation. The botanical excursions are optional and have been converted into sight-seeing or pleasure trips. The submission of herbarium specimens, collections, slides etc. during degree as well as post-graduate examinations has remained only a formality. There are now centers and shops opened in many Indian cities which deal with ready-made herbarium specimens for sale or on hire. The situation is further deteriorating. The doctoral degrees in floristics and taxonomy are being sold in the market. It would therefore be better if we talk least about this situation.

It is earnestly desired that this situation must be improved with immediate effect and with all seriousness. The first step in this direction would be to prepare a well balanced and full-proof syllabus for angiosperm/plant taxonomy and make it compulsory for all students of botany. The syllabus can be framed by Indian Association of Angiosperm Taxonomists (IAAT) and sent to U.G.C. for recommendations. This syllabus should ideally include certain fundamental principles as well as taxonomic procedures. It is expected that this syllabus should inculcate basic studies and interest in the subject amongst undergraduate students. IAAT should conduct refresher courses in different centres for the training of teachers from all over the country. Once this basic frame work is established, the college students as well as researchers should be encouraged to undertake floristic surveys of small or large areas. The careful surveys with personal critical observation of natural variations, ecological notes and local uses should be carried out on a large scale. The preservation of plant material, conservation of rare and endangered species, cultivation of economically important plants by establishing herbaria, museums and botanical gardens should be encouraged. This phase of plant taxonomy, namely the pioneer-phase or exploratory phase should be satisfactorily completed within a relatively short period.

The next phase of consolidation of the data should be immediately undertaken where the pioneer phase has been completed satisfactorily. Careful identification, revision of smaller groups, monographs, floras and manual writing etc. should be the main activities in this phase. The revisions and monographs must necessarily include information on comparative study of anatomy, cytology and ecology. Other disciplines such as palynology, embryology, phytochemistry etc. may also be undertaken wherever possible. This phase may continue for a relatively longer period and might be concluded within 25-30 years.

Ultimately, all these investigations should aim at the detailed biosystematic studies of various problematic genera and small families. The biosystematic phase may continue for a much longer period and a thorough knowledge about the plant wealth of our country is expected to be readily available in this phase. It would be too early to frame a detailed programme of such investigations at this stage. The details can really be worked out as and when necessary.

The phase-wise investigation of plant-wealth of our country has no other alternative. If our agriculture, botany, forestry, horticulture, indigenous medicine and other plant based industries are to progress and prosper, we must have a sound knowledge of our vast natural resources, namely the wild plants. The disciplines such as Ayurveda and activities such as plant breeding, afforestations, gardening etc. can be fruitful only if our knowledge of basic botany is sound.

Lastly, even the basic botany and taxonomy in our country is being ridiculed outside. This is mainly because our plant researches are based on doubtfully determined plant material. Our reports on chromosome numbers, anatomical or embryological peculiarities, new plant diseases, new host range of parasitic fungi, new photosynthetic pathways, substitutes of valuable medicinal drugs etc. are taken with a pinch of salt only because they are not supported by voucher specimens. We all must join hands in wiping out this shameful affair and try to stand on honest and sincere footing. Plant taxonomy, the oldest of the plant sciences, must regain its glory and should flourish further. Our botanists must be able to produce better evolutionary classifications. They should appear at the fountain-head of all kinds of plant sciences. They should be in a position to establish laboratories and plant research centres and attract students even from the developed countries. This is certainly not a day dreaming. What all we need is a sincere, honest and devoted mind and a self respect.

With the establishment of Indian Association for Angiosperm Taxonomy in the recent years, we hope to tread our path in a proper direction and look forward to the glorified plant taxonomy in this land of Parasharas and Sharangdharas.

ACKNOWLEDGEMENTS

I wish to acknowledge my sincere thanks to Dr. M. Daniel, Department of Botany, M.S. University to share my views on Floristics in India.

REFERENCES

Basak, R.K. (1981). *Taxon* **30**:784-93.

Basak, R.K. (1983). *Botanical Survey of India*: *An account of its establishment, developments and activities*. Howrah:BSI.

Bhandari, M.M. (1978). *Flora of the Indian Desert.* Jodhpur, Scientific Publishers, ed. **2**, 1991.

Blatter, E. and C. McCann. (1930-35). *J. Bombay Nat. Hist. Soc.* **34**:291-306; 623-637; 877-900; **36**:524-537; 781-795; **37**:15-35; 255-277; 523-548; 764-779.

Burkill, I.H. (1965). *Chapters on the History of Botany in India*. Calcutta.

Chaudhury, H.P. (1932). *Pres. Addr. XIX Ind. Sci. Congr.* Bangalore.

Cooke, T. (1901-1908). *The Flora of the Presidency of Bombay*. Vol. I & II. London, BSI reprint Vols. I-III. 1958.

Hooker, J.D. (1872-1897). *The Flora of British India* Vols. I-VII. London: L. Reeve.

Hora, S.L. (1953). *Bull. Nat. Inst. Sci. India* **7**:19-24.

Howard, A. (1926). *Presidential address to Indian Science Congress*, Bombay.

Lawrence, G.H.M. (1951). *Taxonomy of Vascular Plants* New York: McMillan.

Maheshwari, J.K. (1963). *The Flora of Delhi*. New Delhi: C.S.I.R.

Majumdar, G.P. (1927). *Vanaspati: Plants and plant Life as in Indian Treatises and Tradition.* Calcutta.

Majumdar, G.P. (1946). *13th All India Oriental Conf. Tech. Sci.* Calcutta. pp. 97-120.

Manilal, K.S. and V.V.Sivarajan (1982). *Flora of Calicut*, Calicut.

Matthew, K.M. (1981-83). *Flora of Tamilnadu Carnatic.* Tirucherapulli: Rapinat Herbarium.

Saldanha, C. and D. Nicolson (1976). *Flora of Hassan District*, Karnataka. New Delhi: Amarind Publ.

Santapau, H. (1953). *Rec. Bot. Surv. Ind.* **16**(1):1-396. Ed.2, 1960; Ed. **3**, 1967.

Santapau, H. (1957). *The Flora of Purandhar*. New Delhi.

Satya Prakash, (1965). *Founders of Science of Ancient India*. New Delhi.

Seal, B.N. (1915): *The Positive Science of the Ancient Hindus*. London.

Sircar, N.N. (1950). *J. Roy. Asiat. Soc. Bengal Letters* **16**:123-39.

Sivarajan, V.V. (1984). *Introduction to the Principles of Plant Taxonomy*. New Delhi: Oxford & IBH Publ. Co., Ed. 2 by N.K.P. Robeson1991. Cambridge: Cambridge Univ. Press.

Stewart, R.R. (1979). *Taxon* **28**:5-12.

Stewart, R.R. (1982). *Taxon* **31**:57-84.

Subramanyam, K. (1979). *Bull. Bot. Surv. India* **19**(1-4):1-4.

Desert Floristics-Need For Updating

M.M. Bhandari and Mahendra Kumar
Department of Botany,
J.N. Vyas University,
Jodhpur-342005, INDIA.

INTRODUCTION

Indian Desert is the eastern extension of the Great Sahara-Sindian Desert and includes the arid and semi-arid regions of the states of Rajasthan, Gujarat, Haryana and Punjab. It is approximately 100 metres above the mean sea level and is situated nearly 300 km. west of **Aravallis** and 600 km. north of Kutch. Most of the area consists of dry undulating plains of hardened sand and the remaining region is a rolling plain of loose sand forming shifting sand-dunes of varying shape, size and form. These dunes may be either simple or complex form of barchan, longitudinal, transverse or parabolic types.

To most people the term Desert gives a picture of a unvegetated undulating sea of wind-swept sand, tree-less landscape, devoid of any sign of life, scorched under the pity-less sun, traversed by camel caravans on endless sandy wastes, with a lonely cactus or thorny bush here or a few mini-groves of date-palm there. While this vision may be true of South American Desert or the Sahara or Arabian Desert, the Thar Desert does not confirm to this pattern of vision. Here we do not have palms or cacti or Oasis to greet the weary caravan clad travellers.

The whole area of Thar Desert is a part of 'Maroosthali', the region of death. A large part of the region is supposed to have remained under the mighty waters of great Indian Ocean. And indeed in many parts of Jaisalmer there still remains the illusion of a receded tide, gone but forgotten to return, leaving large stretches of parched land punctuated by sand dunes in systematised curves.

Evolution of desert vegetation in time

Indian Desert is the extreme 'Paroxysmal' form of the original environment that has evolved by a very slow deterioration. Plant life has been gradually transformed until it has disappeared altogether, contributing to the worsening climatic conditions, and finally creating absolute desert. That Indian Desert was not as it is, in not too distant past, but was covered with thick vegetation is proved by the petrified remains of large tree trunks as are found in the Akal Fossil Park, only a few kilometers away from Jaisalmer.

Floristic studies in Indian Arid Zone

The first general record of the plants of Rajasthan was published by Mr. & Mrs. Allen in 1852 in the book 'Views and flowers from Gujarat and Rajpootana'. Hooker and Thomson (1855) in their 'Introductory essay' in the 'Flora Indica' made few remarks on the vegetation of Rajwara (States of Rajasthan) and observed some resemblance in flora of Jaisalmer and that of southern Punjab. Sir George King, a surgeon in Jodhpur State, who later became Superintendent of Saharanpur Botanical Garden in 1868 and in 1890 established the "Botanical Survey of India" at the Royal Botanic Garden (now Indian Botanic Gardens, Sibpur), Calcutta, was the first to give a scientific account of the vegetation of Rajputana (1869) in his 'Sketch of the Flora of Rajputana'. He also gave an interesting information about the famine foods of Marwar (1879). Of course many of his identifications merited scrutiny.

Brandis (1874) in his 'Forest Flora of North West and Central India' incorporated some observations on plants of Rajputana after touring some parts of the state in 1869-70. An anonymous publication entitled 'Introductory note to Jodhpur and Jaisalmer trees and plants' is believed to have been published by Miss Macadam, presumably in the year 1890. Adams (1899) listed about 50 species of plants in his book 'Western Rajputana States'.

Unlike the vast deserts of North Africa, Arabia, Central Asia and even the New World which have attracted the attention of many botanists, the Indian Desert has been rather overlooked. This lack of interest prompted Blatter and Hallberg (1918-21) to study the vegetation of Indian Desert. The systematic study of Blatter and Hallberg discovered 16 new species and 14 new varieties in the Indian Desert. However, a critical study of these new taxa has revealed that many of them are African elements. Blatter & Hallberg themselves indicated that the better knowledge of Kutch, Sind & Baluchistan flora might reduce the number of endemic species, which they had reported to be new to this area. Many of these species have now been reduced to the synonymy of earlier known African elements. More than seventy years after the work of Blatter and Hallberg (1918-21), interest in the taxonomic study of the plants of the Indian Desert has been both lively and considerable. As a result of the work of several agencies, the situation now is entirely different and the arid zone of India is perhaps one of the best botanically explored regions of India (Shetty & Pandey, 1977). Earlier the botanical explorations were undertaken in some selected localities of western Rajasthan by the Western Circle of Botanical Survey of India, Poona between the years 1957-1960, particularly in the districts of Barmer, Jaisalmer, Jodhpur and Bikaner. Some collections have been made from Ganganagar and Pali districts also. In 1964 botanical collections from Jaisalmer and Jodhpur were made by Central Circle of Botanical Survey, Allahabad. Biswas (1952), Biswas and Rolla (1953), Sarup (1957a), Puri and Jain (1960, 1962), have dealt with the vegetation of Rajasthan, some of them very briefly. An account of the vegetation and flora of Jodhpur division has been given by Rolla and Kanodia (1962, 63). Kanodia and Gupta (1969) dealt with the sand- dune flora of western Rajasthan. Ramchandra (1941) published a list of some plants from Rajasthan, S.W. Punjab, Sind, Baluchistan and Kathiawar. Sarup (1951, 1954, 1957b, 1958a) listed plants of Jodhpur and its neighbourhood Bikaner and Jaisalmer. The vegetation in Ladnu in Nagur district and Kailana in Jodhpur district were studied by Tandon (1968) and Bhimya *et al.* (1964) respectively. Contributions have been made towards the flora of the semi-arid parts of Jhunjhunun district, particularly Pilani, Chirawa, Jhunjhunu and Mandrela and their neighbourhood (Mulay and Ratnam, 1950; Bakshi, 1954; Nair, 1956. Dillon and Bajwa (1972) enumerated the plant of Ganganagar district. Reference to aquatic vegetation or plants has also been made by Sarup (1957c, 1958c, d). Sarup (1957d, 1958b) also made a study of the halophytes of the Indian desert. Recently, Saxena and Gupta (1973) dealt with the vegetation of Pachpadra salt basin.

Similarly some contributions from the Taxonomy Laboratory of Botany, Department of Jodhpur University were published concerning the new plant records and also on the identity and nomenclature of desert plants. New records for westen Rajasthan were also published by Nair & Deshpande (1960), Jain and Kotwal (1960), Raizada and Jain (1961) and Jain (1963). All these contributions have improved our knowledge about the desert plants and resulted in the publication by Puri ***et al.*** (1964) in 'The first Flora of Western Rajasthan'. It provided a key to identification and short descriptions of the plants as well. Unfortunately many errors of misidentification, synonymy and nomenclature, etc. crept in this Flora and it was critically reviewed by Gupta & Bhandari (1965). Since then a series of papers have been published and they have substantially added to our knowledge of plants of Indian Arid Zone. A detailed flora of the Indian Desert including the districts of Barmer, Jaisalmer and Jodhpur was published by Bhandari (1978).

Nair and Deshpande (1960), reported the occurrence of *Acanthospermum hispidum* in Rajasthan. Jain & Kotwal (1960) reported eight species of plants not reported earlier from Western Rajasthan. Bhandari (1961) reported a large number of new records for North-West Indian desert. A detailed note on the same was published by Bhandari (1963) in which nearly 25 new taxa have been reported.

Satyanarayan & Shankarnarayan (1964) described a new species of *Lasiurus* i.e. *L. caudatus* from W. Rajasthan. Formerly this genus was represented by only two species i.e. *L. sindicus* Henr. and *L. hirsutus* (Forsk.) Boiss. However, the main character on which this species is based, is the absence of tail in the lower glumes of the sessile as well as pedicelled spikelets, and which character according to authors, distinguishes it from *L. sindicus*.

Blatter and Hallberg (1918-1921) discovered two new species of *Rhynchosia* from the Indian Desert - i.e. *R. arenaria* and *R. rhombifolia*. A joint paper on the correct identity and nomenclature of these two supposed new species of *Rhynchosia* was published by Bhandari & Verdcourt (1970). These two species are the African elements widely distributed in the Indian Desert.

Dignathia hirtella Stapf, an east African species of grass was reported as new for India by Raizada & Jain (1961). They, however, doubted its occurrence in the drier parts of Rajasthan. In addition to their material from Kutch, they quoted Blatter 4654, 'Kotda, Jadodar', Oct. 1917, E.J. Blatter, near Sea. This sheet was examined. The label on this sheet is not in Blatter's handwriting. Blatter and his party visited Indian Desert in October-November, 1917 and collected a number of plants from Kotda - a village near 'Seu' of the pargana Seu (Sheo) in Barmer District. Clearly this Blatter's specimen is from Jodhpur. The arguments is further strengthened by other related field numbers of Blatter, collected from Kotda, Sheo.

Dactyliandra, a monotypic Cucurbitaceous genus, new to the Indian Flora, was, for the first time, recorded by Bhandari & Singh (1964). This plant is well distributed all over the Indian Desert, but somehow or other it remained unnoticed and unrecorded in any of the Indian Floras. While recording *D. welwitschii* from Indian Desert it was remarked that it apparently was a case of discontinuous distribution for it was earlier reported only from Angola and S.W. Africa (Meeuse, 1962). It was thought that this remarkable geographical disjunction can be resolved by the anticipated wide distribution of the species in the intermediate places and the gap in the distribution could be reduced. In fact, Jeffrey in Kew Bulletin (1971) subsequently describing the second species of *Dactyliandra*, reduced this gap by reporting it from Tropical East Africa.

A taxonomic revision of the genus *Indigofera* in Rajasthan was made by Nair & Koshy (1963). *Sesamum mulianum*, a new species was described by Nair (1963). Bhandari (1964) gave correct names to seven of the Indian Desert taxa. Hingorani & Gaur (1965) reported the occurrence of *Heliotropium curassivicum* L. in Nagaur district of Rajasthan. This is a maritime species and its occurrence in land is quite interesting. This taxon has since then been collected from Pachpadra salt areas. As far as Kutch is concerned, Bhandari (1965) reported seven taxa which are well distributed in Indian Desert and have now extended into Kutch region as well.

Jain & Misra (1965) described two new varieties of *Ricinus communis* from arid zone. Ram Deo (1965), in a note added some new records for Rajasthan. Satyanarayan & Gaur (1966) described a new host of *Cistanche tubulosa* in *Aerva persica*. Verma *et al.* (1965) contributed some additions to the flora of Rajasthan. Jain (1966) reported the occurrence of two less known species of *Cynodon*, in addition to the commonly occurring *Cynodon dactylon (L.)* Pers. from Rajasthan. Bhandari (1967) set right some nomenclature and taxonomic identity of a few plants, reported the occurrence of two species of the genus as additional records from the area. Murti & Srivastava (1969) contributed a paper on the Flora of Rajasthan.

Majumdar (1969,71,75) in a series of papers gave correct names to a few taxa, but most of them are from the Southern and Eastern Rajasthan. *Caralluma edulis* (Edgew). Benth. & Hook, reported earlier only from Baluchistan, Sind, W. Punjab, Multan (all in Pakistan), was reported as a new record for India (Bhandari, 1971).

Some new records from the district of Ganganagar were made by Dhillon & Bhandari (1973). Singh, V. (1975) made some interesting plant records from Rajasthan. Bhandari (1973) gave an account of Flora of Jodhpur. Neotype of *Farsetia macrantha* was indicated by Bhandari (1975). Roy (1976) reviewed the genus *Eragrostis* in Rajasthan. Singh, V. (1976) made a taxonomic study of the genus *Cassia* in Rajasthan.

A new species, *Tribulus rajasthanensis*, had been described by Bhandari and Sharma (1976). The species is widely distributed in Western Rajasthan and Ajmer as well. It is usually confined to rocky habitats. Singh Paramjit (1982) gave an extended distribution of the species.

A new species of *Sida*, *S. tiagii* was described by Bhandari (1977). A species of *Datura* has recently been collected from near Sardarsamand and also from Bikaner and has been confirmed as *Datura ferox* Linn. (Bhandari & Mehta, 1980). This is a new record for India. Roy & Singh (1981) reported *Vicia macrantha* Retz. and *Sporobolus aroides* (Torr.) Jorr., as new record not only for Western Rajasthan but new to Indian Flora.

Natural plant resources of Western Rajasthan were dealt with by Singh & Shetty (1977). Gupta & Kanodia (1968) and Bhandari

1974) gave an account of the indigenous vegetable products which re used as food during times of famines in the Indian Desert.

The floristic composition of Rajasthan was reviewed by Jain 1972) who also gave a brief history of botanical exploration in Rajasthan. Dhillon & Bajwa (1972) published a review on the Botanical Contribution of Ganganagar District. Dhillon and Bhandari (1974) eported 16 species of plants which had not been reported previously rom any part of Rajasthan. These are mostly Mediterranean or Malaysian or temperate Himalayan elements. Rao, S.R. (1979), made ome comments on the Indian Plant Taxonomy. Shetty *et al.* (1980) gave an account of some new plant records for Rajasthan. Singh & Shetty (1982) reported 70 species of plants not reported by Bhandari 1978) in his "*Flora of the Indian Desert*". Parmar & Pandey (1984) made a new combination in *Pulicarea rajputanae* Blatt. & Hallb. Bhandari (1984) made a critical study of genus *Farsetia* Turra. in the ndian Desert. Pandey & Shetty (1984) contributed a paper on the Flora of Pali district. Johri (1984) surveyed genus *Ipomoea* in Rajashan in which he reported 2 species and 2 varieties as new records or Rajasthan. Shetty & Pandey (1984) made botanical observations on some biologically significant areas in Western Rajasthan Desert. Parmar *et al.* (1985) studied the flora of Bikaner district. Pandey *et al.* (1986) contributed to the flora of Chura district.

The pioneering work of Hooker and Thomson (l.c.) dividing ndia into floristic provinces was continued further by Clarke (1896) who included Rajputana in his *Subarea* 'India Deserta' which coincides roughly with the 'Indus Plain Region' of Calder (1937), and Chatterjee (1939). Agharkar (1952) regarded the flora of Rajputana desert as a mixture of western (Arabian-African), eastern (Indo-Malayan) and general (Indian) and widely distributed elements. He also observed that they are found in varying proportions in different parts, the western types more common in the west and the eastern types being found near the foot of the Aravallis. Blatter and Hallberg (1918-21) and Biswas and Rolla (1953) found the eastern elements to be about one-seventh of the western and pointed out that the Indo-Malayan and Western Botanical regions meet in Western Rajputana desert. They, thus, supported Drude's (1890) contention that the line of demarcation between the Indo-Malayan and 'Perso-Arabian' flora ranges from the Gulf of Cambay northwards along the Aravallis. Meher-Homji (1970) and Meher-Homji and Bharucha (1975) analysed the floristic elements according to vegetation types and concluded that the demarcation corresponds in general with that of Drude (1890). A brief account of pattern of plant distribution in Rajasthan was given by Mulay (1960). Meher-Homji (1965) showed that the strength of the north African steppe (Sudano-Deccanian) elements in the northern semi-arid zone is two times higher when compared with that in the entire southern semi-arid zone including the Deccan. Sharma, S. (1965, 1980) also deal with some phytogeographic aspects of Rajasthan. Majumdar (1975) gave a brief account of the origin, nature and economic aspects of Rajasthan Flora.

Floristic composition

Great interest is attached to the genera and especially the families of plants which are confined or even nearly confined to the desert. They are the living representatives of the races of plants which have undergone a great change in entering the desert and in acquiring all the characteristics that have made possible their survival and success in the desert since a long period of time. A study of the desert plants indicate that the evolutionary development of desert vegetation, which has accompanied their entry into the regions of aridity and brilliant sun-shine, has in many cases been of far reaching importance. But in majority of cases the plants which, have adjusted themselves to the desert conditions have only undergone the modification of roots, stems, leaves and other vegetative structures. Some of these changes have been so profound as to make the appearance of these plants wholly unlike that of their nearest relatives of the adjacent more humid areas, and yet their flowers and fruits remain almost unchanged. The characters inherited from ancestors of high moisture requirements have been reduced and transformed by the evolutionary processes which have accompanied the settlements of plants in desert environment.

On the basis of recent revision, the flora of the Indian Desert (Bhandari, 1990) consists of nearly 682 species which are distributed among 351 genera and 87 families of flowering plants. 31 additional varieties have been known. Out of this, 9 families, 37 genera and 63 species are introduced.

Future considerations and need for up-dating

Taking into consideration the published literature, Jain (1972) grouped the districts of western Rajasthan under the following categories:

1. Fairly well explored: Jodhpur and Jhunjhunu districts.
2. Underexplored: Barmer, Jaisalmer, Bikaner, Sikar and Churu districts.
3. Almost unexplored: Jalor, Pali, Nagaur and Ganganagar districts.

However, on the basis of the publication and the botanical exploration undertaken in recent years, particularly by the Botanical Survey of India, the districts in western Rajasthan can be reclassified as follows:

1. Well explored: Jodhpur, Pali, Barmer, Jaisalmer, Bikaner and Churu districts.

2. Fairly well: Jhunjhunu and Ganganagar districts.

3. Underexplored: Sikar and Jalor districts.

4. Almost unexplored: Nagaur district.

The arid zone circle of the Botanical Survey of India was se up in 1972 at Jodhpur for studying the flora of the regions in N.W India which are by and large arid of semi-arid. After its setting up intensive botanical explorations have been undertaken in the district of Barmer, Bikaner, Churu, Pali and Jodhpur, covering almost th entire district in each case. Botanical explorations in Jaisalmer an western parts of Ganganagar districts have also been partly complet ed.

Jain (1972) had suggested the undertaking of monographi works on small families on genera pertaining to Rajasthan, and ha also emphasised that, importance should be given to genera whic are economically important or are more common. Monographic stud ies of this nature have been published for the genus *Indigofera* Linn (Nair and Koshy, l.c.) and also for the genus *Aristida* Linn. (Kanodi and Nanda, l.c.) recently. However, revision of the genus *Cassi* Linn. in Rajasthan (V. Singh, l.c.) and the genus *Eragrostis* P. Beauv in Rajasthan (G.P. Roy, l.c.) have been undertaken at the Arid Zon Circle of Botanical Survey of India, Jodhpur.

While dealing with floristics, emphasis so far has mainly bee laid on Angiospermic vegetation. The knowledge of the lower group of plants, which should be of equal importance in the study of floristics, is yet fragmentary and this has indeed been a neglected area of research. A comprehensive survey of the algal flora is long overdue and the need is especially urgent in view of the role they play in biological nitrogen fixation, photosynthesis range clearance and production of food. The same is true for fungi, lichen, liverworts, mosses, ferns etc. the importance of which, in the production of medicine, dyes, antibiotics, etc. are too well known to be discussed.

When nation thinks of facing problems of the 21st century, it has to evolve qualitative environment for the future generation. Plant wealth which is being lost through rapid loss of natural habitats, especially in India, has a big share and has to be protected. The land-use patterns which are predominantly being influenced by agriculture and industrial production, have to be managed more wisely. However, now a days attention is being focused on marginal lands and waste-lands, which have specific and or extreme problems like drought, desertification, flooding, soil-erosion, soil toxicity and soil salinity. All these problems confront agriculturists, foresters, ecologists, and environmentalists, thus challenging them to search such plant species which can not only reproduce and maintain themselves in the fragile ecosystem or stress environment but are also important ensuring sustainabale utilization of bioresources.

Among the conditions that determine the inadequacy of our knowledge and understanding of the Desert Flora, is the size and extent of the area, difficulties and discomforts of its exploration, problems of its international boundary, poverty of the people of the area, and the destruction of vegetation before many of the plants that make it up have been discovered. Most of the collecting and study of the desert flora has been by botanists who are at best visitors or temporary residents in the desert. Results of these circumstances are several. For many districts which form the Indian Desert, often only annotated lists are available, many of which are woefully out of date. Outstanding exceptions to this have in the past been few, but this is being rapidly changed. With the publication of "*Flora of the Indian Desert*" (Bhandari, 1978, 1990) and "Flora of Rajasthan" by Shetty & Singh (1987-94) the scene is getting changed. The coverage, however, of vast areas, is sparse and anything, but adequate. Of most districts which constitute the Indian Desert, the flora is patchy at best, and in many desert districts important habitats where canal water has been introduced have been largely or entirely cleared and brought under cultivation. Even the districts which have been intensively or extensively been explored, reveal that some species formerly indicated as not rare, have rarely, if ever, been found recently, despite intensive collecting during the past decade.

Western Rajasthan is perhaps one of the most botanically explored region of India due to the work done at the University of Jodhpur (Bhandari, 1978, 1990), the CAZRI and Arid Zone Circle of BSI. Whereas BSI's work is included in 'Flora of Rajasthan' 3 volumes (l.c.), Bhandari's work is mostly confined to the Districts of Barmer, Jaisalmer and Jodhpur. But the Indian Desert extends far beyond these three districts and other districts are either under explored or unexplored. Moreover, near Pakistan border, there are large number of ridges and hilly outcrops, rocky areas which are inaccesible due to number of difficulties, remain unapprochable and consequently remained unexplored.

There is urgent need for updating the Flora of these districts and also the districts of Haryana, Punjab and Gujarat where the desert exists. Further with the introduction of the largest canal of Asia, the Indira Gandhi Canal, the waters have brought down new plants from the hilly regions of Punjab which have now established themselves along the banks of the canal. All these need proper inventorization. Certainly, in view of the rapid degradation of the vegetation of the desert, the most urgent task of near future is carefully planned and thorough collection of germplasm. Our knowledge of plants in their normal habitats and relatively unchanged by human activities, are diminishing daily. The importance of conservation and protection of endangered Flora has been fully realized the world over.

REFERENCES

Adams, A. (1899): *The Western Rajputana States*, Taylor & Francis, London, pp. 397-422.

Agharker, S.P. (1952): *Bull. Nat. Inst. Sci. India* **1**:246-47.

Allen, G.L. (1852): *Views and Flowers from Gujarat and Rajputana, London*, pp 1-15.

Bakshi, T.S. Bombay *Nat. Hist. Soc.* **52**:484-514.

Bhandari, M.M. (1961): *Proc. 49th Ind. Sci. Cong.* **3**:107-108.

Bhandari, M.M. (1963): *Proc. Raj. Acad. Sci.* **10**:40-50.

Bhandari, M.M. (1964): *Ann. Arid Zone* **2**:181-184.

Bhandari, M.M. (1965): *Ann. Arid Zone* **4**·217 229.

Bhandari, M.M. (1967): *Ann. Arid Zone* **6**:200-210.

Bhandari, M.M. (1971): *J. Bombay Nat. Hist. Soc.* **68**:296-297.

Bhandari, M.M. (1973): *Souvr. Vol. Summ. Inst. Sci. Jodhpur,* **1**:8-14.

Bhandari, M.M. (1974): *Economic Botany* **28**:73-81.

Bhandari, M.M. (1975): *J. Bombay Nat. Hist. Soc.* **72**:604-606.

Bhandari, M.M. (1977): *Ann. Arid. Zone* **16**(4):455-458.

Bhandari, M.M. (1978): *Flora of the Indian Desert*, MPS Repros, Jodhpur, pp 1-435.

Bhandari, M.M. (1984): *J. Econ. Tax. Bot.* **5**(5):1087-94.

Bhandari, M.M. (1990): *Flora of the Indian Desert,* MPS Repros, Jodhpur, pp 1-435.

Bhandari, M.M. & D. Singh (1964): *Kew Bull.* **19**:133-138.

Bhandari, M.M. & M. Mehta (1980): *J. Bombay Nat. Hist. Soc.* **76**(3):550-51.

Bhandari, M.M. & V.S. Sharma (1976): *Bot. Not.* **129**:367-369.

Bhandari, M.M. & B. Verdcourt (1972): *Bull. Bot. Surv. India* **12**:108-112.

Bhimaya, C.P., A. Cherian and Y. Satyanarayan (1964): *Indian For.* **90**:667-675.

Biswas, K. (1952): *Bull. Nat. Inst. Sci. India* **1**:247.

Biswas, K. and S.R. Rolla (1953): *Bull. Nat. Inst. Sci. India* **19**B:411-421.

Blatter, E. & F. Hallberg (1918-1921): *J. Bombay Nat. Hist. Soc.* **26**:218-246, 525-551, 811-818, 968-987; **27**:40-47, 270-279, 506-519.

Brandis, D. (1874): *The Forest of Northwest* and *Central India. London*.

Calder, C.C. (1973): *Ind. Sci. Cong. Asso.* 71-91.

Chatterjee, D. (1939): *J. Roy. Asiat. Soc. Bengal* **5**:19-67.

Clarke, C.B. (1898): *J. Linn. Soc. Lond.* **34**:1-146.

Dhillon, K.B.S. and S.P. Bajwa (1972): *Bull. Bot. Surv. India* **11**:234-244.

Dhillon, K.B.S. and M.M. Bhandari (1974): *J. Bom. Nat. Hist. Soc.* **70**:577-582.

Drude, O. (1890): *Handbuch der Pflanzengeographie*. Stutgart.

Gupta, R.K. and K.C. Kanodia (1968): *J. Agri. Trop. Bot. Appl.* **15**:265-285.

Gupta, R.K. and M.M. Bhandari (1965): *Ann. Arid Zone* **4**:236-238.

Hingorani, G.R. & R.C. Gaur (1965): *Indian For.* **91**:293.

Hooker, J.D. & T. Thomson (1855): *Flora Indica*, London, pp 1-247.

Jain, S.K. (1963): *Proc. Nat. Acad. Sci. India* (Sect.B) **33**(1):58-60.

Jain, S.K. (1966): *Indian For.* **92**:201-202.

Jain, S.K. (1972): *Bull. Bot. Surv. India* **12**:176-187.

Jain, S.K. and M.N. Kotwal (1960): *Sci. & Cult.* **26**:191-192.

Jain, T.C. and D.K. Mishra (1965): *Ann. Arid Zone* **4**:43-45.

Johri, S.C. (1984): *J. Eco. Taxo. Bot.* **5**(5):1113-1142.

Kanodia, K.C. & R.K. Gupta (1969): *J. Bomb. Nat. Hist. Soc.* **65**:681-695.

Kanodia, K.C. & P.C. Nanda (1968): *Indian For.* **94**:296-302.

King, G. (1869): *Proc. Asiat. Soc.* **38**:116-122.

King, G. (1879): *Indian For.* **4**:226-236.

Macadam, Miss (1890): *Introductory note to the Jodhpur and Jaisalmer trees and plants.*

Majumdar, R.B. (1969): *Bull. Bot. Soc. Bengal* **23**:35-40.

Majumdar, R.B. (1971): *Bull. Bot. Soc. Bengal* **25**:75-76.

Majumdar, R.B. (1975): *Bull. Bot. Soc. Bengal* **29**:135-140.

Meeuse, A.D.J. (1962): *Bothalia* **8**:10.

Meher-Homji, V.M. (1965): *J. Bom. Nat. Hist. Soc.* **62**(1):15-18.

Meher-Homji, V.M. (1970): *Vegetatio* **21**:299-320.

Meher-Homji, V.M. & F.R. Barucha (1975): In *Environmental Analysis of the Thar Desert* (Gupta, R.K. & I. Prakash, eds.), pp 237-273.

Mulay, B.N. (1960): *Memo. of Ind. Bot. Soc.* **3**:9-12.

Mulay, B.N. & B.V. Ratnam (1950): *Proc. Indian Sci. Cong.* Pt. **3**:64-65.

Murti, S.K. & J.N. Srivastava (1969): *Bull. Bot. Soc. Bengal* (Sen Mem. Vol.), pp 461-468.

Nair, N.C. (1956):*Proc. Raj. Acad. Sci.* **6**:49-64.

Nair, N.C. (1961): *J. Bombay Nat. Hist. Soc.* **58**:433-440.

Nair, N.C. (1963): *Bull. Bot. Surv. India*. **5**:251-253.

Nair, N.C. & M.B. Deshpande (1960): *J. Bombay Nat. Hist. Soc.* **57**:441-442.

Nair, N.C. & T.T. Koshy (1963): *J. Bombay Nat. Hist. Soc.* **60**:326-336.

Pandey, R.P., Parmar, P.J. & B.L. Vyas (1984): *J. Eco. Tax. Bot.* **5**(2):473-476.

Pandey, R.P., Parmar, P.J. & G.P. Roy (1986): *J. Eco. Tax. Bot.* **7**(2):323-349.

Pandey, R.P. & B.V. Shetty (1984): *J. Eco. Tax. Bot.* **5**(2):335-378.

Parmar, P.J. & R.P. Pandey (1984): *J. Eco. Tax. Bot.* **3**:273-278.

Parmar, P.J., Pandey, R.P. & G.P. Roy (1985): *J. Eco. Tax. Bot.* **7**(1):55-67.

Puri, G.S. & S.K. Jain (1960): *Indian For.* **86**:85-86.

Puri, G.S. & S.K. Jain (1962a) *Trop. Ecol.* **3**:49-69.

Puri, G.S., Jain , Mukerjee, S.K., Sarup, S. & N.N. Kotwal (1964): *Rec. Bot. Surv. India* **19**:1-159.

Raizada, M.B. & S.K. Jain (1961): *Indian For.* **87**(7):426-428.

Ramchandra, R.Y. (1941): *ICAR Bull* **43**:1-45.

Ramdeo, K.D. (1965): *Indian For.* **91**:123-125.

Rao, S.R.(1979): *In Tropical Botany* (Larsen K. & L.B. Holm-nitelsen), Ahrus, pp 153-163.

Rolla, S.R. & K.C. Kanodia (1962): *Ann. Arid Zone* **1**:16-46.

Rolla, S.R. & K.C. Kanodia (1963): *Ann. Arid Zone* **2**:36-60.

Roy, G.P. (1976): *Bull. Bot. Surv. India.* **18**:102-108.

Roy, G.P. & V. Singh (1981): *J. Bom. Nat. Hist. Soc.* **73**(3):532-534.

Sarup, S.A. (1951): *Rajputana Univ. Stud.* **1**:29-35.

Sarup, S.A. (1954): *A list of some common plants of Jodhpur & its neighbourhood, Jaipur,* 1-25.

Sarup, S.A. (1957a): *Dungar Coll. Mag.* **18**:1-8.

Sarup, S.A. (1957b): *A list of some common plants of Bikaner & its neighbourhood.* 1-12.

Sarup, S.A. (1957c): *J. Indian Bot. Soc.* **36**:597.

Sarup, S.A. (1957d): *J. Indian Bot. Soc.* **36**:598.

Sarup, S.A. (1958a): *A list of some common plants of Jaisalmer & its neighbourhood.* 1-16.

Sarup, S.A. (1958b): *Univ. Rajputana Stud.* **3**:55-60.

Sarup, S.A. (1958c): *Univ. Rajputana Stud.* **3**:61-70.

Sarup, S.A. (1958d): *Univ. Rajputana Stud.* **3**:71-76.

Satyanarayan, Y. & K.A. Shankarnarayan (1964): *J. Bombay Nat. Hist. Soc.* **60**:763-76.

Satyanarayan, Y. & Y.D. Gaur (1966): *J. Bom. Nat. Soc.* **62**:602.

Saxena, S.K. & R.K. Gupta (1973): *J. Bom. Nat. Hist. Soc.* **70**:104-127.

Sharma, S. (1965): *Univ. Rajasthan Stud.* (Bot.), 1-8.

Sharma, S. (1980): *J. Eco. Tax. Bot.* **1**(1-2):55-75.

Shetty, B.V. & R.P. Pandey (1977): *Bull. Bot. Surv. India* 15.

Shetty, B.V. & R.P. Pandey (1984): *Indo-US Workshop on Arid Zone Res., Jodhpur* 1-12.

Shetty, B.V. & V. Singh (1987-94): *Flora of Rajasthan* (3 Vols.) Bot. Surv. India, Calcutta.

Shetty, B. V., Vyas, B. L., Panday, R.P. & Parmar, P.J. (1980): *Bull. Bot. Surv. India* **22**:72-79.

Singh, P. (1982): *Bull. Bot. Surv. India.* **24**(1-4):237-238.

Singh, V. (1976): *Bull. Bot. Surv. Ind.* **18**:85-101.

Singh, V. & B.B. Shetty (1977): *Trans. Isdt. Ucds.* **2**(2):296-305.

Tandon, S.K. (1968): *The Vegetation of Ladnu* (Jodhpur), Jaipur, 17-26.

Verma, D.M., Wadawa, B. M. & O.P. Mishra (1965): *Proc. Nat. Acad. Sci.* **35**B:163-170.

Species Concepts - Quo Vadis?

M.P. Nayar
The Greens, 19/315, Vatavila,
Trivandrum - 695006.

Species, a basic unit of classification, has conceptually changed over time and space to include all aspects of alpha and omega taxonomy. Species is a dynamic system and it generally follows its genetic code and changes its patterns through speciation to the demands of environmental stresses. Rollins (1957) appropriately said "the field of Taxonomy is in a way epitomizes the work of all other branches of biology centered on the organism itself and brings the varied factual information from them to bear on the problems of relationship, classification and evolution".

The history of Systematic Botany is divided into six major periods: (1) The Gross Morphology period (400 BC - 1700 AD) (2) Internal Morphology period (1700-1858) (3) Evolutionary period (1858-1900) (4) The Cytogenetical period (1900-1960) (5) The chemical taxonomy period (1940-) and (6) Molecular and gene finger printing period (1960-). These periods are evolutionary due to developments of sophisticated instrumentations and technologies available to the taxonomist, like microscopes, electron scanning microscopes, mass spectroscopy, X-ray crystallography and computers. The data that are generated in all phases cumulatively and continuously upgraded the vast corpus of knowledge in Taxonomy. In spite of the vast data, nearly after 2000 years of Taxonomy, it is estimated we knew only by name 1.5 million species of plants and animals compared to the estimated 10 to 15 million species. Unless these are named, their extinction would also remain unknown. It is estimated by Peter Raven (1988) that, of the 180,000 tropical flowering plants, about 135,0000 species would remain in patches of disturbed forests in the beginning of 21st century and they are doomed to extinction during the next 30 years unless remedial steps are taken.

The question being asked today, how much we can do it save them. Taxonomists should not get bogged down in the polemics of nomenclature and classification. Already Taxonomy has reincarnated itself as Biodiversity and there is great opportunity for Taxonomists to play an effective role in indentifying "hot spots" of plant diversity which would help in the conservation of plants.

SPECIES CONCEPTS

There are five major species concepts developed over the years as per the perceptions of different schools of systematists. The platform is littered with acrimonious debates without comprehending the purpose and motive of each school. Though there is underlying common thread of interlinkages, each school tried to belittle the other school as seen in the following debates and in this process Taxonomists forgot to save the plants that sustain their debates!

The major species concepts are (1) Taxonomic species concept (TSC) (2) Biological species concept (BSC) (3) Evolutionary species concept (ESC) (4) Operational species concept (OSC) and (5) Genetic species concept (GSC).

Taxonomic Species Concept (TSC)

The history of taxonomic species concept is Synonymous with the history of botanical studies throughout the world in Herbal gardens for medicine and later on through the vast collection of plants from far and nearby plant collectors accompanying the merchant navigators and colonisers in the unchartered continents and seas in the period of colonialism of the world by few countries of Europe during 15th to 18th century. This was a period of great confusion in the identity of species and polynomial name did not help much. Linnaeus, a Swedish Botanist came to the arena with binomial nomenclature and artificial sexual system of classification. It was part of the history of the sorting out the vast array of herbarium plants and pickled specimens in some order and form resulting in a heirarchial system as per subjective perceptions which include sorting them out in different clusters: families, genera and species. Of course morphology was the main guiding force and this resulted in a system of nomenclature. These later resulted in the International Code of Botanical Nomenclature first held in Paris in 1867 and later refined and revised in several International Botanical Congresses resulting at present 14th, International Code of Botanical Nomenclature. These Codes brought out the Type concept, the rules of naming different categories, families, genera and species and the rules of priority and procedures of publications. The moving up and moving down of taxa from one category to another as per taxonomic monographic studies and rules of priority brought in several name changes which became a controversial one for the school of experimental biological taxonomists. Formalising the nomenclature and where overlapping characters do not find particular slot created suspicion among experimental taxonomists. The subjective opinion of taxonomists is questioned on the basis of few experimental studies. As a practical step, the herbarium identification of plants is the simple method of retrieval of plant identity through standard specimens properly authenticated by an expert. It does not give legality for the view of the fixity of species, since Nomenclature Type is a constituent element of a taxon and is not necessarily the most representative one. The value of taxonomic species concept enhances proportionately to the number of characters used and the catholicity of outlook in the soul of a good taxonomist. Since Linnaeus introduced the binomial nomenclature, Taxonomic concept with its rules of nomenclature stood the test of time, as no other systems, however scientific they are, could not tackle the vast array of plants. Taxonomic species based on the temporal phenotypic discontinuous variations, serves as a frame work to which one can add or modify as per modern scientific inputs. It is a practical retrieval of identity through Herbarium type specimen. The rules of the nomenclature instability is partly solved through *Genera conservanda* and *Species conservanda*. A practicing taxonomist uses his years of experience as a tool for alloting categories and this subjective decision is akin to the judgements in jurisprudence by judges of eminence. There would be no opportunity for experimental taxonomists to test taxonomic categories in a vaccum. Since there is a vast array of plants and animals which require identification and since alpha taxonomists are professionally downgraded by vociferous experimentalists, there is a great dearth of Taxonomists in the world for different groups of plants and animals. The frame-work or hypothesis would no longer be available for experimentalists as taxonomists world over have become an endangered category. Taxonomy and biodiversity have become prominent in this era of man-made extinction of species. It is even proposed by several eminent Taxonomists to have a band of "para taxonomists" for recording the species before they become extinct.

Philosophy of Intuition and creativity

All great discoveries in science became possible first by human imagination and creativity. This was later proved in the crucible of experimentation. To say alpha taxonomists are considered subjective, is tantamount to discounting human subjective creativity. Appropriately Lincoln constance (1964) mentioned. "Speculation, imagination, intuition are all precious attributes of human mind and must not be proscribed. Without them even the most severely quantitative science would never have gotten off the ground. Since in my view systematics is an art as well as science, these phases of the process of understanding should be welcomed into the general synthesis". C.P. Snow, well-known scholar who wrote "Two cultures" would have welcomed this aspect as according to him natural science and humanities are part of the holistic development of human mind.

Biological Species Concept (BSC)

Alpha taxonomists define separate taxa where there are significant morphological gaps which coincide upto the boundaries of interbreeding. The biological species is the fundamental unit of organisation in biparental organisms. It is a reproductively isolated system of breeding population. (Grant, 1981). In terms of biological species, the ability to exchange genes and to cross, belong to the same biological species. Where reproductive barriers develop which stop gene exchange, there is boundary for biological species. When it comes to sibling species with no morphological differences, but show reproductive barriers it is seen that the traditional taxonomic species concept breaks down. Heslop Harrison (1954) considers, formal taxonomy a ready tool and the term "species" is applied to the traditional taxonomic hierarchy and advocated the use of the term "hologamademe" for the Biological species concept. However Grant (1957) proposed the term "binom" for "Species" of Taxonomic species concept and the term 'Species' to be given for the Biological species concept.

The Biological species concept was already visualised by prelinnean biologists like Ray (1686) when he said that a species name is to be given the name only when it breeds true. This was ignored by Linnaeus (1751) as he had to contend with the large number of herbarium specimens reaching him from different parts of the world. Nearly 150 years elapsed since Taxonomists realised the meaning of biological species concepts of Muntzing (1930) and Love

(1967). That is, genepools isolating from one another resulted in speciation *i.e.* chromosomal differences in populations. Pierre Ledgendre (1972) used mathematical tools in formalising definitions of biological species. A species is defined "as a group of individuals which cluster when one applies on them relation showing their possibility of crossing freely." Grant (1981) mentioned that the biological species is "the sum total of interbreeding population". A local population is defined as a group of organisms located within pairing distance of each other (Legendre, 1972). Ghiselin (1987) and Hull (1976) questioned whether species are really individuals, but observed that they could be termed philosophically as individuals in spatio-temporal canvas. While Mayr (1942) proposed that biological species is "a group of actually or potentially interbreeding populations, reproductively isolated from other such groups". The lines of approach or process of thinking for Taxonomic species concepts and Biological species concepts are in two different planes and they run in parallel tracts. This is best explained by Heslop Harrison (1954) and Sonneborn (1957) as they elucidated that TSC is for the purpose of classification and for easy retrieval of information, and the other (BSC) is for the purpose of finding out barriers of genepools of biological species. Hence Heslop Harrison appropriately preferred to use the term "species" for Taxonomic Species concepts and the other new term "hologamodeme" for the Biological species concepts. (interbreeding group of populations). The taxonomic species concept is practical and useful, when dealing with the large array of species and it gives a framework. The Biological species concept comprehends sibling species which are phenotypically same but genotypically different, while taxonomists could only include sibling species in the taxonomic heirarchy.

Biological species concept has ingradients of evolutionary approach as Nature's creative force in the genetical system. According to Judith Groch (1969), "if life's duplication process were to produce exact replicas of the present generation, life could not have evolved. Nature is neither an assembly line nor a museum; life is neither static or repetitious. The canvas of life is an evolving masterpiece which creator is itself."...."Life, an enormous deck of cards, chromosomes and genes are constantly shuffled and reshuffled making possible an astronomical number of new combinations". It is the process of natural selection which is the orienting force in evolution giving order and form to the unprocessed novelties born of nature's fertile "imagination". Modern theories of evolution maintain that the environment in providing the challenge to which a species must respond becomes the directive force in the evolutionary processes". The biological species concept is concerned with biparental organisms. In order to bring out the synthesis of uniparental and biparental organisms which are part of life systems, there is a need for a common platform for comprehending uniparental and biparental organisms.

Evolutionary Species Concepts (ESC)

The evolutionary species concept has arisen out of this perceived idea of uniting the uniparental and biparental biosystems. Simpson's (1951, 1961) evolutionary species concept is on the basis of the population which has the following characteristics: It is an ancestral descendant population and it maintains its identity from other lineages and it has definite "unitary evolutionary role" and definite evolutionary tendencies in space and time, concomitant with natural selection which is opportunistic for a particular ecological niche of a biotic community.

Phylogenetic Species Concept (PSC)

Darwin's "Origin of Species" has an evolutionary base which perceived that all organisms have descended from one another in a linear fashion. Hayata (1931) proposed network of relationship between species which means that evolution need not follow in a linear fashion and this was supported by Lam (1936). However Donoghue (1985) and Mishler and Brandon (1987) supported that the grouping is linear and monophyletic. Inspite of the occurrence of parallel evolution, taxonomists like Thorne (1963), Alston & Turner (1963) stated that phylogeny gives meaning and substance to the taxonomic categories. Thorne (1963) states that "evolutionary studies add meaning, stature and excitement to our science" Walters (1961) and Sokhal (1963) warn that taxonomists should not enter into the web of conjectures on phylogeny, when the data base on fossils or evolutionary mechanisms are meagre. Lincoln Constance (1964) argues that if taxonomists cannot enter into the field of evolutionary thinking, no other science can enter into the field of evolution as information obtained by biosystematists are scanty for any meaningful discussion.

Convergence or parallel evolution in unrelated groups is well explained by Went (1971) *i.e.* (i) occurrence of cushion plants in Alpine mountains in the Himalayas occurring in different families: *Leguminosae, Saxifragaceae, Caryophyllaceae, Asteraceae;* (ii) Leaf convergence in Mediterranean climates (iii) Eucalypt leaf types in Australia occurring in 12 families (iv) divaricate shrubs in New Zealand with 51 species in 23 families.

Contributions of Cytology, Chemical and Numerical Taxonomy

Ehrendorfer (1964) admirably summed up the contributions of cytology, but still there are large gaps in the knowledge system of plant evolution. There is good information system on chemical taxonomy by the study of betacyanins versus true anthocyanins in Centrospermae and of sulphur compounds in Rhoeodales. It is seen in major groups, phytochemistry has been studied haphazardly and there is no meaningful picture of comparative data for systematic application. Alston *et al.* (1963) stated that "at present time there is no phylogenetic system based on the distribution of biochemical Classification,".

Sokhal & Sneath (1963), while advocating numerical taxonomy, stated that "the practice of taxonomy remained intuitive and commonly inarticulate, an art rather than science". The computer taximetrics which is based on non weighted character of all characters of plant is hailed Neo-Adansonian thinking in clustering and repeatability of studies are appropriately assessed. It is emphasised that "the primary aims of phenetic species concepts are repeatability and objectivity". However Cracraft (1987) refined Sokhal's views and stated that operational species concepts (OPC) include species as clusters of organisms which are basically distinguishable from operational units in cladistic analysis. However Thorne (1963) advocated that giving equal taxonomic weight to every character is a backward step as it does not take into consideration evolution of characters and conservative characters. However the modern computer data bank is a great source of strength to the taxonomists. (Electronic Data Processing - EDP).

Genetic Concept of Species

The genetic basis of species is well explained by Grant (1981) where multiple gene systems are seen in *Nicotiana longifolia* and chromosomal rearrangements in *Brassica, Clarkria* and *Nicotiana*. Clausen et al., (1945) said "Natural species consist of individuals whose genes are in internal balance so that a harmonious physiologic development is assured generation after generation". The usual ways of speciation are gradual speciation and rapid speciation.

The gradual speciation occurs over a span of time and space in different environmental parameters faced by the species resulting in the formation of number of races, *i.e.* species like *Agrostis* occurring in dump heaps of heavy metals or *Astragalus* species occurring in physiologically intolerable alpine regions. Rapid speciation takes place through polyploidy which are usually allopolyploids, through hybridization between parental species followed by doubling of chromosomes or through rearrangement of the genes in the chromosomes. Another mode of speciation is through the sexual reproduction in apomictic groups and the progeny becoming genetically different from the parent and again such genetically different individuals get stabilised or perpetuated by subsequent apomixis. e.g: *Rubus, Taraxacum*.

Conclusion

Crick's Prophesy of protein taxonomy is possible through DNA isozyme analysis now. Silby (1962) has, through analysis of bird's egg proteins, unraveled the arrangement of their constituent amino acids which is directly related to gene structure. It is also possible to measure homologies among the gene through DNA hybridization (Hoyer, *et al.*, 1964). It is accepted that the double stranded DNA helix is possible through DNA of closely similar or related organisms. DNA finger printing and ultimate classification of biological groups and lineages can be established through analysis of nucleic acids and enzyme proteins. Ehrendorfer (1987) indicated that "generally evolutionary success or failure depends on a very complex interplay of not clearly separable organismal and environmental parameters". Ultimately, Taxonomy is an unending synthesis of all our knowledge systems.

LITERATURE CITED

Alston, R.E., & B.L. Turner. (1963). *Biochemical systematics:* Prentice Hall: Englewood Cliffs: N.J. xii + 401 pp.

Alston, R.E., Mabry, T.R. & B.L. Tuner. (1963): *Science*. **142**: 545-552.

Clausen, J., Keck, D.D. & W.M. Hiesey. (1945) *Carnegie Inst. Washington Publ.* 564.

Constance Lincoln (1964): *Taxon* **13**(8):257-273.

Donoghue, M.J. (1985): *The Bryologist* **88**:172-181.

Ehrendorfer, F. (1964): *Vista Botany* **4**:99-186.

Ehrendorfer, F. (1987): In *Proceed. of xiv International Botanical Congress*; ed. W. Gruener & B. Zimmer, 317-333. Berlin.

Grant, V. (1981): *Plant Speciation* Ed. 2, Columbia University Press, New York.

Hayata, B. (1931): *Bot. Gaz.* **49**:328-348.

Hoyer, B.H., McCarthy, B.J. & E.T. Bolten. (1964): *Science* **144**:959-967.

Judith Groch. (1969): *"The Right to create"*: 1-393, Little, Brown & Company, Boston & Toronto.

Lam, H.J. (1936): *Acta Biotheoretica* **AZ** 133-194.

Linnaeus, C. (1951): *Philosophica botanica*. Stockholmiae 1-362 pp.

Love, A & D. Love. (1967): *Grad. Bot. Contrib. Bot.* 203:210.

Mayr, E. (1942): *Systematics and the origin of species*: Columbia University Press, New York.

Mischeler, B.D. and M.J. Donoghue. (1982): *Syst. Zool.* **31**:491-503.

Muntzing, A. (1930): *Hereditas*: **13**:185-341.

Pierre Ledgendre. (1972): *Taxon.* **21**:381-406.

Raven, P.H. (1988): *Taxon*: 549-560.

Ray, J. (1686): *Historia Plantarum vol. 1*. Mariae Clark, Londinii: 1-983. pp.

Rollins, R.C. (1957): *Amer. Journ. Bot.* **44**:188-196.

Simpson, G.G. (1951): *Evolution* **5**:285-298.

Simpson, G.G. (1961): *Principles of Animal Taxonomy*. Columbia.

Silby, C.G. (1962): *Syst. Zool.* **11**:108-118.

Sokhal, R.R. (1963): *Taxon*. **12**:190-199.

Sokhal, R.R. & P.H.A. Sneath (1963): *Principles of numerical Taxonomy*. Freeman, San Francisco & London XVI + 359 pp.

Sonneborn, T.M. (1957): In: *The species problem*: ed. by E. Mayr. Amer. Assoc. Adv. Sc Washington DC.

Thorne, R.F. (1963): *Amer. Nat.* **97**:287-305.

Went, F.W. (1971): *Taxon.* **20**(1):197-226. 1971.

Botanical Nomenclature

N. Ravi
Verbena, Mundakkal Middle, Kollam, 691001.

In the early period of plant taxonomy *i.e.* before the middle of the 18th century there were no definite rules governing plant nomenclature. During that period the plant names were *Polynomials*. Then the plants were known by descriptive phrases *i.e.*, a plant name consisted of a series of words constituting a more or less terse or brief description of the plants in Latin. For example, the plant which is commonly called Bulbous Buttercup and botanically called *Ranunculus bulbosus* was known during the period of polynominals as *Ranunculus foliis, Ovatis serratis, Scapo nudo unifloro* - a brief description of the plant in the Latin *i.e.* Ranunculus plant with ovate serrate leaf, having a scape carrying a single flower without any bracts. Similarly *Sida cordifolia* was named *Althaea maderaspathana, Subrotundo folio molli* et *hirsuto multiplis* by Plukenet in 1692.

With the discovery of more and more taxa all over the world this cumbersome polynomial system ran into difficulties or rather it became quite inoperative. Thus in 1753, Carolus Linnaeus put forward in his *Species plantarum* the Binomial System of Nomenclature which very easily replaced the polynomial system.

According to the Binomial System of Nomenclature the name of a plant species consisted of two epithets or words. The first denoted the name of the genus to which the species belonged and the second the specific epithet. These two words together formed a binomial e.g. *Crotalaria striata, Hibiscus rosa-sinensis* etc. The significance of this Linnaean contribution can be assessed from the fact that this system has been incorporated in the present-day rules of nomenclature *in toto* without any modification or amendments. Not only that, this binomial system has been accepted even by present-day zoologists.

The adoption of the binomial system alone could not prevent the confusions that were to arise in the early period of botanical nomenclature with the discovery of more and more taxa all over the world. In that early period there was not any international body or system to coordinate the work of the taxonomists spread all over the world. Consequently, the taxonomists of the various countries worked independently and as they liked. This had resulted in much confusion in the field of nomenclature.

If we trace the early history of plant nomenclature, multiplicity of plant and language difficulty were only two of the several problems which confronted the taxonomists of that period.

As the time passed the need for a uniform procedure for naming and describing plants arose. A movement was started towards this end. The Swiss botanist Alfonse De Candolle, the celebrated son of Augustine Pyrame De Candolle headed this movement. His efforts resulted in the first international Botanical Congress held in Paris in August 1867. Invitations were sent to 150 European and American botanists to attend the Congress. In advance of their arrival, De Candolle sent to each of the invitees, for preliminary study, a copy of his "Lois de la nomenclature botanique" (Laws of botanical nomenclature). They were subjected to discussion and finally it was resolved that "these laws, as adopted by this assembly, shall be recommended as the best Guide for Nomenclature in the Vegetable Kingdom". These rules comprised the first Code of Nomenclature called *Paris Code, 1867*. The rules of this Code were often known as "de Candolle Rules".

The *Paris Code,* 1867, represented an excellent beginning in the right direction. But as the time passed, the application of the Paris Code revealed numerous inherent deficiencies. The need for revision and modification became apparent with the result that scarcely a decade had elapsed before various schools of botanical thought were putting into practice individual interpretations and "rules" that were not a part of the Paris Code. This has resulted in a split in the botanical fraternity. Many American and European botanists and some others started practicing a new set of rules called "Kew Rules".

In the United States, a group of botanists under the leadership of Nathaniel Lord Britton of the New York Botanical Garden, developed a set of rules to govern plant nomenclature. They were presented at the meeting of the Botanical Club of the American Association for the advancement of Science held in Rochester, New York in 1892. These rules constituted the *Rochester Code* 1892.

The staunch advocation of these rules and their employment by the advocates of the *Rochester Code* created two opposing schools of thought among American botanists. One group under the leadership of N.L. Britton supported Rochester Code and the other group under the leadership of Asa Gray supported Kew Rules.

In 1900, a convention of botanists was held in Paris. This was probably the second International Congress of Botanists. At this convention, plans were made for a serious and detailed study of the entire nomenclatural problem. This study lasted for five years and culminated in the third International Congress held in Vienna in June, 1905. At this conference, the advocates of the Rochester Code tried unsuccessfully to incorporate the rules of the Rochester Code in the new code being formulated. The code which came into being at this

Congress was a modified and amplified version of the Paris Code 1867. The advocates of the Rochester Code refused to subscribe to the *Vienna Code* 1905. For this reason the Vienna Code did not gain world-wide acceptance.

Meanwhile, the supporters of the Rochester Code, established a new code of nomenclature called *American Code 1907*. Though this code was an extension of the Rochester Code, it agreed with the Vienna Code in some respects. It was the first sign of rapproachment between the two opposing schools of botanical thought. Yet its sponsorship and circle of adherents remained essentially unchanged and was mainly that, which supported the Rochester Code movement.

Until 1930 differences of opinion persisted among the differing groups. The 1930 Cambridge (England) Congress saw signs of harmony and accord returning to the international botanical fraternity. At the Cambridge Congress, every effort was made to harmonize the basic differences between the Vienna Code and American Code. The credit for bringing the two groups together should go to T.A. Sprague and M.L. Green. As a result, the new rules legislated at Cambridge were a product of conciliation and accord by parties of both factions. Thus for the first time in botanical history, a code of nomenclature came into being that was international in function as well as in name. It was the first International Code of Botanical Nomenclature (abbreviated as ICBN). This was the *Cambridge Code* 1930.

After this, several international botanical congresses were held in different parts of the world. At each of these Congresses, amendments were made in the Code of the previous Congress. Thus after the *Cambridge Code* came *Amsterdam Code 1947*, *Stockolm Code 1952, Paris Code 1956, Montreal Code 1961, Edinburgh Code*

1966, Seattle Code 1972, Leningrand Code 1978, Sydney Code 1981, Berlin Code 1986, and finally *Tokyo Code 1991*.

What is the Code? It is just like the constitution of a country. As the people swear by the Constitution of their country the botanists have to swear by the Code of Botanical Nomenclature. It is as sacrosanct as the constitution of a country.

Just as the constitution of a country, the code also has a Preamble, Divisions, Chapters and Appendices.

The Preamble means preface or introduction. The Division 1 contains the Principles which do not give detailed rules about nomenclature but show the main ideas that have guided the compilers of the Code. Division II contains Rules on Articles which are 75 in number and also Recommendations. The 75 articles are included under IV Chapters. Division III contains provisions for future modification of the Code. Besides these, there are 3 appendices.

Division 1 contains altogether 6 Principles which form the basis of Botanical Nomenclature. They are as follows:

1. Botanical nomenclature is independent of zoological nomenclature (For the purpose of this Code, the word 'plants' does not include bacteria).

2. The application of names of taxonomic groups is determined by means of nomenclatural types.

3. The nomenclature of a taxonomic group is based upon priority of publication.

4. Each taxonomic group with a particular circumscription, position, and rank can bear only one correct name, the earliest that is in accordance with the Rules, except in specified cases.

5. Scientific names of taxonomic groups are treatea as Latin regardless of their derivation.

6. The rules of nomenclature are retroactive unless expressly limited.

The Rules (Articles) and Recommendations contained in Division II are the following: Articles, which are 75 in number, are the "laws" of nomenclature. They are mandatory on all botanists who accept the Code. That is, a botanist who accepts the Code has to follow the articles in their letter and spirit, or in other words, he has to follow them blindly. But the Recommendations, on the other hand, are provided for guidance; they are only recommendatory in nature. They have been provided to bring about greater uniformity and clarity especially in future. They are often practical applications of the rules. Though they are not mandatory, the convention is that they are also adopted as such by botanists. The rules and recommendations apply to all organisms treated as plants (including fungi), whether recent or fossil, with the exception of the bacteria.

Division III contains the provisions regulating the modification of the Code. This is intended for future botanists who would attempt to amend the Code.

The Appendices deal with names of hybrids, *Nomina familiarum conservanda, Nomina generica conservanda et rejicienda, Nomina specifica Conservanda et rejicienda* etc.

The present note does not intend to make a detailed study of the Code. Only the important points are discussed here.

One of the Principles of the Code says that names of taxonomic groups are treated as Latin regardless of their origin. The name of every taxon is a Latin name whether derived from the Greek or any other language. In other words, plant names are all latin words or latinized version of any other language.

E.g. *Digitalis* - a latin word meaning fox-glove.

Lithospermum - latinized form of two Greek words, Litho = stone and sperma = seed.

Caesalpinia - a latinized name after the Italian botanist and physician, Andrea caesalpino who is referred to as the first plant taxonomist.

Bauhinia - is a latinized version of a name given after two French-Swiss botanist brothers, Gaspard (Gasper) Bauhin & Jean (Johann) Bauhin.

Lindera - Latinized version of a name given after a Swedish Physician John Linder.

One of the early articles of the Code states that, names of species should be binomials. A binomial consists of two epithets, the generic and specific epithets. e.g. *Crotalaria striata*. The specific epithet may indicate one of many features. It may indicate the country or place where it is abundantly seen, or a country or place from where it was first reported, or it may be after a famous personality, botanical or political (national) or in most cases, it will indicate a very constant character of the species.

E.g. *Mangifera indica* - an Indian plant.
Hevea braziliensis - a Brazilian plant
Oenothera lamarkiana - after Lamarck
Pilea microphylla - small leaves
Sida cordifolia - cordate leaf

It should be understood how a binomial should be written or printed. There should always be a space separating the generic epithet from the specific. It is mandatory that the generic epithet is spelled with a capital initial letter while the specific epithet is ordinarily spelled with a small initial letter. Formerly a capital inital letter was used for certain specific epithets which were derived from the names of persons or dieties, vernacular names etc.

E.g. *Griffithella Hookeriana*

In this connection there is a Recommendation *viz.*, all specific and trivial names should be written with a small initial letter, although writers desiring to use capital initial letter for particular names or epithets may do so when these are derived directly from the names of persons (dieties) or are venacular (or barbaric) names, or are previously published (including pre Linnaean and invalid) unmodified generic names.

This Recommendation advocates the decapitalization of the initial letter of all specific and infraspecific epithets. At the same time, it recognizes the right or prerogative of those who capitalize initial letter of specific names when taken from the names of persons (*Griffithella Hookeriana*, in honour of Sir. J.D. Hooker, *Pinus Coulteri*, after William Coulter), when taken from another genus (*Dianthus Caryophyllus*, the epithet *Caryophyllus* being a pre Linnaean generic name or when taken from a vernacular or barbaric name (*Schinus Molle*, the epithet Molle being a Peruvian vernacular name). But the present convention is to do away with capitalization of the initial letter of a specific or infraspecific epithet irrespective of its origin.

There are Articles which deal with author citation and Recommendations concerning it. Accordingly, the name of a taxon (unitary, binary or ternary) should be followed by the name/names of author/authors who first published the name concerned.

e.g. *Annonaceae* Jussieu
Annona Linnaeus
Annona squamosa Linneaeus
Cyprepedium calceolus Linnaeus var. *pubescens* (Willdenow) Correll.

This is done to make the name more accurate and complete. This, of course, can be done only in scientific papers, Journals, Manuals, Floras etc. The author's names put after the names of plants are abbreviated unless they are very short. One syllable and short names, in general, are not abbreviated or abridged. When long, they are abbreviated; usually the first consonants only are given e.g. Fr. for Fries. When the names are abbreviated care is taken that they are clear and understandable.

E.g. L. or Linn. for Linnaeus, DC for A.P. De Candolle, H.B.K. for Humboldt, Bonpland & Kunth, R.Br. for Robert Brown and W.Br. for William Brown.

After the abbreviation a(.) is a must to indicate that it is an abbreviation. With the sons of several famous botanists like Linnaeus & Hooker coming to the scene, in order to avoid confusion and to distinguish the father botanist from the son botanist the Latin term 'filius' (meaning son) is used abbreviated as 'f'. and given the abbreviated name of the father e.g. Linn. means father Linnaeus and Linn.f. means son Linnaeus. Similarly Hook.f. stood for Sir J.D. Hooker son of the botanist William Hooker whose name is abbreviated as Hook.

Here it has to be understood that author's name is not a part of the name of the plant. To differentiate them it is customary to indicate the scientific name of the plant in print by italic types. It has also become a custom that in hand-written and type-written matters the scientific names of species and of taxa of lower categories are underlined. This is done to distinguish them from the other letters.

Type Method (Typification)

There are several articles in the present Code which deal with Type Method. Type Method stipulates that the name of a taxon should be on the basis of a type called *nomenclatural type*. The type of an Order should be the typical Family belonging to it. The type of a family is its typical Genus, the type of a genus is its typical species. The type of a species, subspecies or variety is a typical specimen belonging to the taxon concerned, made into a herbarium specimen or other element. In the cases of higher taxa viz., Order, Family, Genus etc. mere mentioning of the type in the paper published will satisfy the provisions concerning type concept. But, in the case of species, subspecies and varieties, the type specimen made should be deposited in a recognised herbarium and mention of that should be made in the paper subsequently published.

The type method was the bone of contention between the warring groups of botanists in the early period of nomenclature. The type concept was evolved by N.L. Britton and his followers in their Rochester Code 1892. This got universal recognition only in 1930 at Cambridge.

Earlier Codes, for example *Stockholm Code* 1952, stipulated the application of the Type Method only to taxa of the rank of species and below. That is, only species, sub-species and varieties came under the purview of the concerned provision. Later the scope of the Type Concept was widened to include taxa of the rank of families and of lower levels. In the present Code the principle of Typification is applicable to names of all taxa of the rank of families and below and to those names of taxa of higher levels which are derived from generic names i.e., at present varietal names, specific names, generic names, family names & those names of higher taxa derived from generic names should be on the basis of a Type.

It should be understood that the type of an order is the typical family of that order. e.g. Magnoliaceae is the type family of the order Magnoliales; Malvaceae is the type family of the Malvales. The order Guttiferales need not have a type according to the Code; because, the name Guttiferales is not based on a generic name. (Now the order Guttiferales have been renamed Clusiales and its type is Clusiaceae (Guttiferae). The type of a family is the typical genus of that particular family. e.g. *Magnolia* is the type genus of Magnoliaeae. Similarly its typical species, *Magnolia virginiana* L. is the type species of the genus *Magnolia*. The nomenclatural type of a species is any specimen (typical) of that species. It is usually a herbarium specimen. Some materials do not lend themselves to preservation of characters in dried forms such as succulents. In such a case, the herbarium sheet of parts that can be so preserved (as flowers) may be augmented by critical and diagnostic drawings and photographs. Some species in the absence of herbarium material may be typified by a description or a figure. Great care is given to type specimens. They are the record for the future as well as the present. Should there be disagreement between the author's description or diagnosis of a plant and characters of that plant as shown by the type specimen, those of the specimen are considered to be correct. Sometimes, a taxonomist, by oversight or carelessness, may describe a plant incorrectly. This incorrectness which has crept into the published description can be rectified later, only if the type specimen of the taxon is available. Great is the value attached to the type specimens that in many institutions they are preserved in special envelopes, folders, cases or vaults.

Nomenclatural types are of different types. They are *Holotypes, Isotypes, Syntypes, Paratypes, Lectotypes* and *Neotypes*.

Holotype: A holotype is the one specimen or other element used by the author of the name or designated by him as the nomenclatural type, i.e., holotype is the one herbarium sheet to which the name of the taxon is permanently attached.

The term can be elaborated with help of an example. Suppose, a taxonomist comes across with a species of *Sida* which is suspected to be new. After examining a large number of specimens collected from different places he confirms the identity of the plant with the help of literature or with the help of experts on *Sida* as a new taxon. He will give it a name according to the rules of nomenclature stipulated in the ICBN and publish that name in a recognized Scientific Journal.

Suppose, he first collected it from the S.N. College Campus and prepared herbarium sheets of it with the Nos. 450 A.C. Before naming it and publishing the name he would collect specimens of that species from all available sources and make them into herbarium sheets. In this way he prepared several herbarium sheets of that species like 450 A.C., collected from the S.N.C. Campus, Quilon, 1250 A.C., collected from Kundara, 2500 A.C. collected from Punalur and 2750 A-C., collected from Trivandrum. All these were subjected to detailed examination to gather all the characters of that plant. Only after examining all these plants which grew in different localities under different climatic conditions, he would prepare the description or diagnosis of the plant. He would first prepare a description in English and then with the help of a Latin Dictionary or with the help of someone who knows Latin, he would prepare a Latin description.

Thus, he is already having collections made from 4 different places at different times. Sometimes all these collections may be typical agreeing fully with the diagnosis of the species. In such a case, he would designate any one specimen from any of the four collections as the Holotype. If all collections are not typical, a specimen from a collection which is most typical will be designated holotype.

Thus it is seen that the *Holotype* is the one specimen or other element, used by the author of the name or designated by him as the nomenclatural type, i.e., it is a specimen or element to which the name of the plant is permanently attached. *The holotype is chosen by the author from a single gathering made by the author or any other collector at one time, and expressed definitely at the time of publication.*

Isotype: Isotype is a duplicate of holotype. In the example described above, 2750 A was designated holotype, the other two specimens of the same collection viz., 2750 B and 2750 C are the isotypes. If there are ten specimens collected from a single spot at one time, one of them is made the holotype and the remaining nine are the isotypes.

Syntype: is any one of two or more specimens used by the author in lieu of a holotype, or any one of two or more specimens simultaneously designated as type.

Suppose, no mention of holotypes is made in the publication, some other specimens are made mention of e.g., 50 A-C collected from S.N.C. Campus, Quilon, 1250 A - C collected from Kundara, 2500 A-C collected from Punalur and 2750 A-C collected from Trivandrum. Any one of these can be designated as syntype when no holotype was designated by the author. Or, when several specimens are simultaneously cited as types, any one of them will be a syntype.

Paratypes: They are specimens cited in the protologue or original description other than the holotype, isotype or syntypes. If a new taxon's description is based on more than one gathering made at different times, the author will designate one gathering as holotype and isotypes. The specimens belonging to the other gatherings bearing different field numbers form paratypes.

In the example explained above, 2750 A was made holotype and 2750 B and C, thus became isotypes. If mention of the other specimens collected from the three other places also is made in the original description, these specimens, viz., 450 A-C, 1250 A-C and 2500 A-C, become paratypes.

Lectotypes: Next in importance to holotype, Lectotype is a specimen or other element selected from the original materials to serve as a nomenclatural type, when no holotype was designated at the time of publication or as long as it is missing. If a holotype is lost or destroyed, the lectotype is chosen from the isotypes. If no holotype was designated and if syntypes exist, one of them must be chosen as lectotype. If neither an isotype, nor a syntype is existing, a paratype if it exists, must be chosen as lectotype.

For e.g. Rangachari & Tadulingam (*Journal of Bombay Natural History Society* **24**:846, 1916) while describing the new species *Cynodon barberi*, cited several specimens in the original publication, but failed to indicate one as the type or holotype. Hence all the specimens cited became syntypes. Jain (*Bulletin of the Botanical Survey of India* **8**:204, 1967), after a critical study, correctly selected one of the specimens - "Madras Herbarium, South India, Flora No. 13715", and deposited it in CAL (Central National Herbarium, Calcutta), as the Lectotype.

Neotype: is a specimen or other element selected to serve as nomenclatural type as long as all the materials on which name of the taxon was based are missing. A neotype may be designated only when all the originally cited materials are believed lost or destroyed. In selecting a neotype more care and critical knowledge are essential as the reviewer usually has no guide except his own judgement as to what best fits the protologue.

E.g., The new species *Farsetia macrantha* (Cruciferae) was described by Blatter & Hallberg (*J.B.N.H.S.*, **26**:220, 1918). In the protologue they cited three numbers viz., Blatter 5758, 7300 and 7305. These specimens were preseumably deposited in the Blatter Herbarium, Bombay, where the entire collection of Blatter including the 'types', has been preserved. Bhandari, J. (*J.B.N.H.S.*, **72**: 606 1976) while working on this species could not trace the type specimens in the Blatter Herbarium. He concluded that all the original materials were either lost of destroyed. Therefore, he designated Bhandari 507 as Neotype, "examplifying the true application of the name *Farsetia macrantha* and to a certain with certainty the sense in which the same must be used. The specimen comes from the type locality and is in complete agreement with the original diagnosis".

Publications

The publication of names is a subject that is strictly legislated by the Code. Whenever a new taxon is named or the name of a known taxon is changed, it has to be published. The present Code contains several articles which suggest where and how the names are to be published.

There are two aspects for the publication of scientific names: *effective* and *valid* publications. Effective publication refers to the place and scope of publication. Until January 1, 1953, a name could be published anywhere in a book or magazine, on a printed herbarium label of a widely distributed herbarium collection, in a newspaper, or in a nursery seed list. The rules of the present Code reject such publications and names published in such a way. The present rule is that, to be effective the name must be published in a book or periodical that is distributed in botanical centres throughout the world in some indelible form. Suppose the name is announced at a public meeting or published in a newspaper which has world-wide circulation, or the publication is by issue of microfilms or otherwise unpublished manuscripts, such a publication is never an *effective publication.*

What is a valid publication? It required that, in addition to being effectively published, the name must meet one or other of two conditions. One is that the name should be accompanied by the Latin description of the taxon. Another is that it must be accompanied by a direct or indirect reference to a previously validly published description. If one of these conditions is observed, the publication becomes valid.

The first condition is self-explanatory. The second is less obvious and therefore requires some explanation.

In 1772, Scopoli separated the species *Panicum sanguinale* Linn. from the genus *Panicum* and put it in another genus *Digitaria* with the name *Digitaria sanguinalis* and published that name in his

2nd edition of the Flora Carniolica on page 52 of Vol. 1, in 1772. Scopoli did not give its Latin diagnosis or description. But he did make a reference to a previous validly published description of the plant as given by Linnaeus under the generic name of *Panicum* in the latter's *Species Plantarum* on page 57 in 1753. Scopoli published the new name in this way:

Digitaria sanguinalis (L.) Scop.
Panicum sanguinale L., SP. 57, 1753.

In this case, the previous name referred to is called *Basionym*. Here the publication is valid since it was published in a Scientific Journal. Though the new name is not accompanied by the Latin description, a reference to the source of the original publication is given. That is sufficient to meet the requirements of the Rule. A very important point to be borne in mind this connection is that a simple reference to a previous validly published name will not satisfy the provisions of the Code. The basionym should be followed by a full reference as to its author and original publication with page or plate reference and date.

Thus for a publication to become valid, if the published name is of a new taxon, the name of the plant, accompanied by its Latin description, should be published in a Scientific Journal that will reach the various botanical centres. When the publication involves change of names, instead of the Latin diagnosis, basionym with full reference should accompany the new name. When none of these conditions is satisfied the publication becomes invalid. Such names are invalid names.

The term invalid does not find a place in the Code, but the names which are not valid are called invalid by botanists. At present, there is a recommendation attached to the Rules concerning valid publication. According to this, publication of the name of a new taxon of the rank of family and below, on or after 1-1-1958, is valid only when the nomenclatural type is indicated in the published paper. The place where the type specimens is permanently preserved also may be indicated in the paper.

In this connection there are two more terms viz., *Legitimate name* and *Illegitimate name. Legitimate name* is one which is in accordance with all the provisions of the Code concerning publication. A name which contravenes any one or more Articles of the Code concerning publication of names is an *illegitimate name*.

Are all valid names legitimate names?

From the definition given earlier it may appear that all valid names are legitimate or they are interchangeable or synonyms. That is not so. There is a distinction between the two. All valid names need not be legitimate names, but a legitimate name is always a valid name. To make the distinction between the two clearer requires more elaboration with the help of another very important Rule of the Code called *Rule of Priority.*

The Rule of Priority is considered the corner stone of the Code. The Articles, concerning the Rule of Priority, state that each taxon from the rank of family and downwards has only one correct name and that is the earliest validly published name. A taxon can have several valid names; only one of them is correct and that is the earliest validly published name.

That the correct name of a taxon is the first valid name is a provision of the Code. A legitimate name is one which is in accordance with all the provisions of the Code. A valid name is one which has been published accompanied by its Latin description or a reference to a previously validly published description. It will not be in accordance with all the provisions, if it does not conform the provision regarding priority. What the Rule of Priority states is that a taxon from the rank of family and downwards has one correct name and that is the earliest valid name. It is the legitimate name. From this it is clear that a taxon can have only one legitimate name and its all other names, valid or invalid, are illegitimate.

The following problem will illustrate clearly the distinctions among these various terms. There is the Apocynal genus *Vinca*, so named by Linnaeus. In 1828, Reichenback split the genus *Vinca* into two. One retained the generic name *Vinca* according to one of the Rules of the Code and the other was given a new name *Lochnera*. Thus he created a new genus *Lochnera* out of the *Vinca* complex. He effectively published the name in a Scientific Journal. But the name was not accompanied by its Latin description. In 1837, George Don renamed *Lochnera* as *Catharanthus* and published that name effectively accompanied by its Latin description. Without knowing this, in August 1838, Endlicher validated the name *Lochnera* by giving its Latin description in a Journal.

What is the legitimate name of the new genus? Which are the valid, invalid and illegitimate names of the taxon?

In 1828, *Lochnera* was invalid. In 1837, *Catharanthus* was valid; *Lochnera* was invalid. In August 1838, *Lochnera* became valid, i.e. after August 1838 the new genus has two valid names, *Catharanthus* and *Lochnera*. Now the Rule of Priority comes into operation. According to it, *Catharanthus* is the Legitimate or correct name and *Lochnera* is illegitimate but valid.

The other valid name of a correct name is called a synonym. The synonym is abbreviated as Syn., therefore, when the name *Catharanthus* is written in a scientific paper it will be a written like this.

Catharanthus, G.Don, 1837.

Syn. *Lochnera*, Reich. ex Endl. 1838.

Here the names of two authors are linked by the Latin word 'ex' meaning 'from'. It means that the name '*Lochnera*' was published validly by Endlicher after taking it from Reichenback.

The principle of Priority is not mandatory for names of taxa above the rank of family. It was originally intended to give stability for plant names, but its strict application causes serious instability in plant nomenclature as and when a botanist unearths some old and forgotten valid publications.

Some explanation is required regarding the meaning of a correct name. The legitimate or correct name is the earliest validly published name. There are instances when a legitimate name becomes correct or incorrect depending on one's concept of a particular taxon. One botanist's concept of a particular taxon may be different from that of another botanist.

E.g., the generic name *Hibiscus* L. of *Species Plantarum* 1753 is a legitimate name in accordance with the rules. Medicus in 1787 created a new genus *Abelmoschus* out of Linnaeus's *Hibiscus* separating the common Lady's finger and other species (spp.) from it. Just as *Hibiscus* is a legitimate name *Abelmoschus* also is a legitimate name. Both generic names are correct when the two genera are kept separate. In 1824, Augustine De Candolle united *Abelmoschus* Medicus and *Hibiscus* Linn. in a single genus *Hibiscus* Linn. When this treatment is accepted, as was followed in most of the earlier Indian Floras, *Hibiscus* Linn. is the only correct name with this particular circumscription. Subsequently, however, Schumann in 1895 and following him several recent taxonomists re-established the genus *Abelmoschus* Medicus. The legitimate name *Abelmoschus* Medicus may therefore be correct or incorrect according to the different concepts of the users. The same is the case of the several new taxa created out of the previous original taxa.

Nomina Conservanda

The Latin words *nomina conservanda* mean preserved or conserved names (sing. nomen conservandum) which are preserved by the Code. There are both family and generic names which are not subject to the Rules of the Code. The Code specifies such names which can be used even against the provisions of the Code. They are the conserved names or *nomina conservanda*. They have been conserved due to their very long usage in botanical nomenclature and in the interest of nomenclatural stability.

There are both family names and generic names which have been conserved in the Code. The family names so conserved are called *Nomina Familiarum Conservanda,* while the conserved generic names are called *Nomina Generica Conservanda*. The concept of *Nomina Conservanda* is more strictly applicable to generic names than to family names.

According to the Code a family name should be after its type genus, e.g., Magnoliaceae after *Magnolia,* Annonaceae after *Annona,* Ranunculaceae after *Ranunculus* and Dipterocarpaceae after *Diptero-carpus* and so on. If this rule is strictly followed, some eight family names become incorrect, e.g., Palmae (Arecaceae), Gramineae (Poaceae), Cruciferae (Brassicaceae), Papilionaceae (Fabaceae), Guttiferae (Clusiaceae), Umbelliferae (Apiaceae), Labiatae (Lamiaceae) and Compositae (Asteraceae). Because of the very long usage in Botanical Nomenclature, these names have been conserved *i.e.,* they have become *Nomina Familiarum Conservanda*. Though these names have been conserved, an alternate name has been suggested for each family (Given in parenthesis). That means, one or the other can be used as the botanist likes.

But in the case of generic names, only those names that have been conserved can be used.

The Papilionaceous genus *Desmodium* was validly published in 1813. The same genus was earlier given valid names *Meibomia* in 1759 and *Pleurolobus* in 1812. According to the Rule of Priority *Meibomia* 1759 has priority over both *Pleurolobus* 1812 and *Desmodium* 1813 and *Pleurolobus* has priority over *Desmodium*. But *Desmodium* has a very long usage while other names have been obsolete. Therefore, against the Rule of Priority, *Desmodium* has been conserved while others have been rejected. Such conserved generic names are called *Nomina Generica Conservanda* and rejected generic names are called *Nomina Generica Rejicienda*.

At the Cambridge Congress (1930) there were suggestions from several taxonomists for conserving certain specific names *i.e. Nomina Specifica Conservanda*. But due to strong objections they were not accepted. But the present Code has accepted the concept of *nomina conservanda* for species also.

In the Code, there are several important nomenclatural terms, some of which need explanation. Such terms are *Autonym, Homonym, Basionym, Combinatio nova, Tautonym, Nomen Nudum* and *Cotype*.

Autonym: is a name automatically established for a taxon, merely as a consequence of the proposal of another taxon. e.g., the recognition of a subspecies or variety that does not contain the type of the species automatically establishes a subspecies or variety that does contain the type. Here the name of the subspecies or variety that contains the type specimen of the species will be an autonym. The final epithet of an autonym, at the infraspecific level, is that of the specific name unaltered but without citation of an author's name.

When a species is split into two subspecies, one of these will be fully agreeing with the original description (protologue) of the species or its type specimen, while the other will be a variant. In this case, the type of the first subspecies will be that of the original species. Therefore, its subspecific name should be the same as that of the specific name. Such a name which is automatically established for a taxon is an autonym. An autonym should not be accompanied by author citation. But the second subspecies is a new taxon. Therefore, it requires a new name; hence a type specimen and a new description and the new subspecific name should be accompanied by the name of the author, *i.e.,* the name of the person, who splits the species and establishes a new taxon. Publication of the name *Ipomoea pes-caprae* (L.) Swt. subsp. *braziliensis* (L.) Ostr., resulting from the splitting of the species *Ipomoea pes-caprae* (L.) Swt., automatically establishes another subspecies, the type of which is that of the original species. Hence its subspecific name is the same as the specific name without author citation. Here the name of the latter subspecies is an autonym.

Ipomoea pes-caprae (L.) Swt. subsp. *pes-caprae*

Ipomoea pes-caprae (L.) Swt. subsp. *braziliensis* (L.) Ostr.

Here the first name is an autonym.

Publication in 1965 of the name *Crotalaria willdenowiana* DC. subspecies *glabrifoliata* Ellis, automatically established another subspecies, *C. willdenowiana* DC. subsp. *willdenowiana,* the type of which is that of *C. willdenowiana* DC.

Homonym: is a name spelled exactly like a validly published name for a taxon of the same rank based on different types. When two distinct taxa of the same rank have the same name spelled exactly similar, such names are called homonyms.

E.g., the name *Jatropha heterophylla* was given by Steudel in 1840 on the basis of a type specimen to a particular species. The same name was given to another species of *Jatropha* by Hooker in his *Flora of British India* (Vol.5) in 1887. They are having separate type specimens and hence separate species. Such identical names are called homonyms. According to the Code vide Rule of Priority, the latter homonym is illegitimate, i.e. *Jatropha heterophylla* Hook.f., is illegitimate or incorrect. Finding out this mistake, Balakrishnan, renamed it as *J. heynei* in 1961.

Tautonym: is a binary name in which the specific epithet exactly repeats the genric one.

E.g. *Malus malus*
Strobus strobus
Pinus pinus

These are tautonyms. According to the Code they are illegitimate or incorrect. Such names as *Cajanus cajan* and *Bambusa bambos* are not tautonyms as the specific epithets do not repeat exactly the generic ones.

Basionym: is the original legitimate name on which another name is typified and published.

e.g. *Digitaria sanguinalis* (Linn.) Scopli was originally *Panicum sanguinale* Linn. Scopoli separated this particular species from *Panicum* and put it in *Digitaria*. According to the Rules of the Code he can do it on taxonomic grounds but he should retain the original specific epithet and indicate the original author's name in parenthesis (brackets). Thus the new name is written like this,

Digitaria sanguinalis (Linn.) Scopoli

Syn. *Panicum sanguinale* Linn.

Here the latter is a basionym.

Combinatio Nova (Comb. Nov.)

It is a new name formed from a previously published name or epithet and is typified by the type of the basionym.

When a taxon of lower rank than genus is transferred to another genus or species, on sound taxonomic grounds, with or without alteration of rank, the original legitimate name or epithet is retained and the author of the basionym is given in parenthesis before the name of the person who effected the alteration.

When *Panicum sanguinale* L. was transferred to the genus *Digitaria* by Scopoli, he published the new name as follows.

Digitaria sanguinalis (L.) Scopoli *Comb. nov.* In this case the abbreviation *comb. nov.* indicates that the new name is a new combination. In this case, there is no alteration in the rank of the new name.

Another example, *Miliusa tomentosa* (Roxb.) Sinclair is a new combination based on *Uvaria tomentosa* Roxb.

If a taxon is relegated to a lower rank, e.g. a species is reduced to the rank of a subspecies, the new name under the subspecific rank will be a new combination. The new combination is followed by the abbreviation *stat. nov.* meaning new status.

E.g.,the new species *Abutilon albescens* Miquel, according to Borrs, did not have characters sufficient enough for treatment as a species distinct from *A. indicum* (L.) Sweet. Therefore, Borrs reduced *A. albescens* to the level of a subspecies of *A. indicum*. Here the new name which is new combination is *Abutilon indicum* (L.) Swt. subsp. *albescens* (Miq.) Borrs *stat. nov.*

Cotype: is a term formerly used in botanical nomenclature for syntype and sometimes erroneously for isotype and paratypes.

Nomen Nudum: is a naked name, *i.e.* name unaccompanied by description, diagram or reference to a previously published description. When the name *Lochnera* was originally published by Reichenback in 1828, it was a *nomen nudum* i.e. *Lochnera* Reich. 1828 is a *nomen nudum*.

REFERENCES

Henry, A.N. & M. Chandra Bose. (1980): *An Aid to the International Code of Botanical Nomenclature*.

Lawrence, G.H.M. (1955): *An Introduction to Plant Taxonomy*.

Lawrence, G.H.M. (1964): *Taxonomy of Vascular Plants*.

Chemical Adaptations of Angiosperms and Their Relevance in Taxonomy

M. Daniel

Phytochemistry Laboratory, Department of Botany, The M.S University of Baroda, Vadodara, 390 002, India.

INTRODUCTION

Angiosperms, the most dominant plant group on the planet earth, are peculiar in that they elaborate a plethora of organic compounds generally known as the secondary metabolites. This group of compounds were much despised in the past and were called 'waste products', 'excretory products' or even named as 'luxury diversifications' by Heslop Harrison. But , of late, the evidences pouring in, have established that the secondary products are the major survival mechanisms of higher plants. These compounds find great use in the struggle existence, in which every individual has to rage a war against a host of enemies such as fungi, bacteria, viruses, insects, herbivores and neighbouring plants. The secondary metabolizes act as antimicrobials, (including phytoalexins), insect-feeding deterrents, bitter principles to ward off herbivores, pollinator guides or allelochemics. They have a decisive role in dormancy, vegetation patterning, pollination and seed dispersal. Time is not far away when we have to accept that the higher plants owe their survived and success to their secondary products. Those plants which produce better chemicals survived best, proliferated and became the dominant taxa on earth while those plants having less efficient chemicals evolved slowly or failed to evolve further.

The compounds which helped the plants on survival are the following: antimicrobials to resist the pathogens, toxins to deter the insects and larger herbivores, pollinator guide to guarantee effective pollination, allelochemics to control the plants of the neighbourhood and to protect the plant from a highly oxidising environment.

A careful scrutiny of the distribution of these compounds in plant kingdom reveals that every plant group produces a characteristic group of compounds or an assortment of compounds. The production of these compounds is genetically controlled. With the help of these compounds the plants struggled for survival. Since the requirements of different environments are different, the plants experimented with various sets of compounds and thus many lines of evolution appeared. In this paper the efficacy of various compounds in contributing to the survival value of plants is critically examined.

1. Antimicrobials

The higher plants produce a wide variety of antibacterial and antifungal compounds. These compounds protect the plant from the seedling stage to the fruiting stage. The water soluble antimicrobial compounds leach out in to the soil at the time of seed germination and sterilize the surrounding environment so that the tender radical and plumule are not attacked. The young leaves, flowers and fruits possess various compounds which protect the plants from microbes, so also the roots, bark and the wood. The main classes of antimicrobial compounds are the following :

a) Phenolics

Phenolics are the aromatic compounds with hydroxyl substituents. It is these compounds which helped the plants to migrate into land from water and are first formed by the delamination of the aminoacids. The evolution of phenols resulted in many classes of compounds such as phenolic acids, coumarins, phenyl propanes, stilbenes, xanthones, flavonoids, tannins, lignin etc. They protect the plants from UV radiation and function as accessory pigments which absorb energy from lower wavelengths of electromagnetic radiation and transfer this energy to other pigments. Flavones were the first polyphenols appeared on earth. They are seen first in Charales and proliferated in liverworts. The ferns elaborated antifungal phenols like catechol, phloroglucionol, resorcinol and pyrogallol. Tannins and lignin also appeared in Pteridophytes. Lignin helped the plant to stand erect for better sunlight. Gymnosperms produced abundant tannins in Pteridosperms. But the higher amount of tannins in these plants inhibited the metabolism of the very same plants and therefore these plants could not survive for long. Biflavones and stibenes were the alternative polyphenols developed by conifers. Angiosperms elaborated various types of polyphenols such as chalcones, aurones, neoflavonoids etc. The primitive Angiosperms contain more flavonoids and other hydroxylated polyphenols while the advanced Angiosperms produced more stable 6-oxygenated flavones and methoxyflavones. It is explicit that various plant groups experimented with different groups of phenols for their survival.

(b) **Volatile oils**

The components of the volatile oils, the monoterpenes and sesquiterpenes, are markedly antimicrobial in nature and were the compounds of choice by conifers. These plants developed diterpenes in resins which also, being antibacterial, increased the resistance of plants to microbial attack. But the volatile oils are found to be less efficient in a tropical climate where they evaporated easily. Therefore only a few Angiosperms developed volatile oils in leaves or stems. They are the Rutaceae, Geraniaceae, Apiaceae, Lamiaceae, and Asteraceae. The Asteraceae volatile oils are peculiar in containing sesquiterpene lactones which act as allelochemics also.

(c) Sulphur containing compounds

The sulphides of the Liliaceae, mustard oil glycosides (glucosinolates) of the Brassicaceae and thiophenes of the Asteraceae are the various plant products containing sulphur. All these compounds are distinctly antimicrobial and confer distinct advantage to plants containing them.

(d) Iridoids

Iridoids are the monoterpenoid glycosides elaborated by the advanced Angiosperms of the Asteridae. They exhibit excellent antimicrobial properties even in very small concentrations. These compounds being bitter in taste, serve as antifeedants also. Iridoids are water soluble and are easily transported. The families which resorted to these compounds, in a big way, are the Loganiaceae, Rubiaceae, Buddlejaceae, Gesneriaceae, Scrophulariaceae and Verbenaceae.

(e) Phytoalexins

Phytoalexins are low molecular weight antibiotics produced by the plants in response to microbial attack. To some extent they constitute the immune-system of the plants responsible for their resistance to microbes. Phytoalexin can be induced by inoculating the plant with a non-pathogenic microbe, stress conditions or even some organic inducers. It is now understood that immediately after its entry into the host, the pathogen releases certain elicitor molecules (oligosaccharins) from their walls due the activity of ß-1, 3-endo glucanase of the host. These elicitors then interact with the complementary receptors on the plant plasma membrane. This interaction generates second messengers which transmit the signal to the nucleus where *de novo* transcription is invoked. The resulting new messenger RNA leads to the synthesis of enzymes involved in phytoalexin biosynthesis and the phytoalexin is formed (Yoshikawa, 1978).

Since many pathogens are specific to the hosts and the phytoalexins do not possess activity against a wide spectrum of pathogens, it is mandatory that every plant group needs to make its own type of compound to combat with its pathogens. The known data on the distribution of phytoalexins agree with this contention. Thus the different plant families tend to accumulate their distinctive class of phytoalexins. For example the Fabaceae, in general, produce is of lavonoid phytoalexins, Solanaceae, diterpenoids; the Asteraceae, polyacetylenes; the Orchidaceae, dihydrophenanthrenes and so on. Within the family also individual genera/species differ from each other in producing different members of the specific class of compounds. The 35 species of *Trigonella* screened for phytoalexins fall into the three groups: the first group with medicarpin and maackiainin, the second group medicarpin alone and third group with medicarpin and vestitol (Ingham and Harborne, 1976).

The result of phytoalexin induction also are found useful in tribal classification of the Fabaceae.

2. **Toxins, Antifeedants and Bitter principles**

These compounds, which form the defence mechanisms against higher animals, are elaborated by plants in large variety. In higher concentrations they may prove toxic to the plants themselves since many of these chemicals are antimetabolites. Alkaloids, terpenoids, tannins and other nitrogen-containing compounds are prominent among them.

(a) **Alkaloids**

Plants experimented with alkaloids in a very large scale and this is evidenced by the very great number of alkaloids produced by them. In fact, the largest group of natural products is alkaloids with the number of known members exceeding 6000. They are basic plant products with a heterocyclic nitrogen and marked pharmacological activity. The different plant groups elaborated a wide variety of alkaloids differing in complexity, structure and basicity to suit their requirements. Yet only about 15% of the Angiosperms produce these compounds.

Alkaloids, being basic, occur as salts of organic acids and thus change the ionic balance within the cell. Some of them act as growth inhibitors (Maisuryan, 1956) while others have varied roles in dormancy (Mastsushima, *et al.*, 1973). The different groups of alkaloids appeared at various times in course of evolution. The primitive angiosperms - Menispermaceae, Ranunculaceae, Paparveraceae and Fumariaceae elaborated complex isoquinoline alkaloids. Rutaceae produced a wide spectrum of alkaloids including quinolines, indoles, simple isoquinolines, tropanes etc. The Fabales produced lupinane alkaloids and the Gentianales, complex terpenoid indole alkaloids. The convolvulaceae developed ergot-alkaloids and the Solonaceae, tropanes. The emphasis on alkaloids as the defensive compounds is lost in the advanced dicots where they are replaced by less toxic compounds.

(b) **Terpenoids**

The higher terpenoids include all the di-tri-and polyterpenoids. Diterpenes, which form the principal compounds of Gymnosperm resins, are bitter in taste and antimicrobial in properties. The toxic principles of the Gentianaceae, such as amarogentin, are some of the bitterest substances in the plant world. The curcurbitacins of the Curcurbitaceae and the nortriterpenoids of the Rutales, including the limonoids and quassinoids, are very bitter in taste and are defensive in function. Phytoecdysones - the insect moulting hormones - are prevalent in ferns as also in Centrospermae. Saponins and cardiac glycosides form another group of toxic principles elaborated by a variety of Angiosperm families. The polyterpenoids such as rubber and gutta contained in the latex of members of Euphorbiaceae, Asclepiadaceae, Sapotaceae etc. also are feeding deterrents.

(c) Cyanogenic glycosides

These simple glycosides, liberating HCN on mild acid hydrolysis, are highly toxic tc herbivores. They are the characteristic compounds of the Rosaceae though they are present in many quite unrelated plants.

(d) Non-protein amino acids

Of the 250 or more non-protein amino acids found in Angiosperms, about one half are present in the Fabaceae. These compounds are multifunctional. Most of them are bitter and provide protection from insects and herivores. They also serve as antimetabolites producing enzymes with reduced catalytic activity and impaired function. A number of them act as allelochemics also.

(f) Tannins

Tannins in the leaves and other plant parts produce a drying sensation in the tongue of herbivores and make the plants containing them tasteless. They also inactivate the digestive enzymes. These compounds are mostly associated with woody habit and are lost in course of evolution. Tannins serve as antimicrobials and protect seeds, fruits and leaves containing them from pathogens. The ability to bind with proteins (enzymes) to make insoluble leather reduces the metabolic rate of the organs containing them. Though in primitive land plants, tannins were systemic and stored in the metabolizing tissues, in advanced plants they are restricted to certain organs like bark, fruit wall, seed coat etc.

3. Pollinator guides

The evolution from anemophily to entomophily necessitated the development of specific pollinator guides which guaranteed that a particular insect visitor goes only to one type of flower. They are color, fragrance, nectar and shape of the flower.

Since the insects and certain birds can 'see' the flowers in the UV region, a number of flowers depended on UV-absorbing flavonoids. A dominant flavonoid or an assortment of flavonoids produce a color which is visible to a pollinator, the eyes of which are sensitive to that region of UV radiation. Other colors developed similarly to attract the pollinators which are sensitive to the visible region of electromagnetic radiations. This is the reason why the flower color varies from white (mostly absorbing UV) to seven colors of spectrum or their intermediate colors. In the visible region the pigment used thus are the anthocyanins, anthochlor pigments and carotenoids. Each of the above-mentioned groups contains compounds absorbing at various wavelengths of a region (color) and therefore every taxon selected a compound/color to suit its requirements. For example, the Polemoniaceae species pollinated by humming birds contain pelargonidin with some cyanidin, lepidopteran-pollinated species contain mainly cyanidin, and bee-pollinated species, mainly delphinidin in flowers. The basic colors are often modified by various acylations or glycosylations. The colorless flavonoids (flavones and flavonols) act as co-pigments of anthocyanins and are required in most flowers for a full expression of anthocyanin color at slightly acidic pH of the cell sap of corolla tissues. Also, when present in sufficient concentrations, they may shift purple or red anthocyanin towards the blue region so that the blue color is often due to a combination of anthocyanin and flavonols in the petals. They are responsible for the cream, buff or ivory colors.

(d) **Volatile compounds**

The importance of volatile oils on flowers is well-known to be discussed here. The dominant monoterpene or a combination of monoterpenes decides the fragrance of a flower. Apart from the mono/sesquiterpene compounds which emit agreeable fragrance, certain flowers produce some volatiles of disagreeable odour for inviting a specific pollinator to the flowers. A few of these compounds are actually the pheromones of the insects pollinating them. For e.g. the aliphatic volatiles (hepadecane, methyl oleate) in some of the Orchidaceae (*Ophrys*) flowers are important attractants to male *Andrena* and *Eucena* spp. bees which pollinate the bee-shaped flowers during pseudo-copulation (Bergstron, 1978). The various species of *Magnolia* have different profiles of hydrocarbons (pentadecane, acetophenone etc.) to attract specific beetles for pollination (Thein *et al.*, 1975). The pheromones benzylacetate eugenol and methylcinnamate of euglossine bees are produced by some orchids of Central and South America which use the bees as pollinators. The unpleasant offensive aminoid odours due to indolic compounds in certain flowers of *Hellebore* and *Iris* form a chemical mimicry by which the plants produce a smell of decaying faces to deceive carrion and dung flies.

4. Allelochemics

In the struggle for existence, allelochemics play a very great role. This aspect of plant-plant interaction is poorly understood even today. Allelochemics are those chemicals which exert a positive influence on nearby plants. Armed with an array of allelochemics a plant in a locality can pattern the vegetation around it just the same way a gardener plans a landscape depending on his whims and fancies.

The known allelochemics fall into a wide spectrum of organic compounds. They are hydroxamic acids in oats (Perez, 1990), phenolic acids and quinines in sedges, sesquitepene lactones in members of Asteraceae (Kelsey and Locken, 1987), non- protein aminoacids in Fabaceae (Chou and Kuo, 1985), coumarins in Apiaceae (Nitao, 1088) and sterols in *Typha* (Greca *et al.*, 1990). The data on the inhibitory effects of plant extracts on other plants are voluminous but in majority of the cases the allelochemics are yet to be identified.

Allelochemics are found to affect every stage of plant growth. They inhibit germination of seeds, reduce root and shoot elongation on seedlings, reduce growth and biomass of mature plants and inhibit flowering and fruiting. The pollen falling on the stigma of another flower inhibits the pollen germination of self pollen or causes the drying of stigma. The decaying plant material of one plant may harbour microbes pathogenic to another species. At times a living plant may serve as a host to the pathogen of a nearby plant.

The allelochemics promise to be an exciting avenue for chemotaxonomy. Since allelochemics deal with the chemical arsenal for offense in a plant, its survival values and evolution is decided greatly by these compounds. Till today a systematic survey on allelochemics has not been carried out and we are still in dark as to the various compounds elaborated by plants to win the race.

5. Antioxidants

The plants produce a wide variety of phenolic compounds. These compounds are seen everywhere in plants. The most important function of these compounds seems to be their action as antioxidants protecting the membrane and lipidic compounds from getting oxidized in the highly oxidising tropical environment. The primitive Angiosperms produced abundant polyhydroxy polyphenols such as myricetin, quercetin, gallic acid etc. These compounds, which were stored in cell sap, get easily oxidized in air and decompose. Therefore in advanced Angiosperms their methoxylated derivatives are produced. The Asteraceae and Lamiaceae, two most advanced families contain highly methoxylated flavonols and flavones. The methoxylation changes the solubility properties in that they become lipophilic and thus occur in the lipid fraction of cells. This seems to be a progressive step to ensure the protection of cell lipids from destruction by free radicals. Another favourable outcome of this modification is that the phenolics can be distributed in surface layers *i.e.*, waxes and cutins, which would screen the UV radiation and combat the pathogen more effectively. The production of quinines by a number of members of the Asteridae such as the Rubiaceae, Boraginaceae, Scrophulariaceae, Orobanchaceae and Verbenaceae supplements the above-mentioned contention (Daniel and Sabnis, 1992).

Conclusions

It is now conclusively proved that the higher plants owe their survival and success to the chemical components they elaborate. It is a sad fact that we are mostly ignorant on the chemical complexity of plants. A systematic screening of plants for their chemical arsenal is the need of the hour. Since a group evolved as a whole it is possible that all the members of a group have similar survival mechanisms. A study of these chemicals will establish the members of a natural taxon conclusively and produce *phylogenetic* groups. The study of defensive as well as offensive mechanisms of all the plants would have a tremendous impacts on the studies of Ecology, Forestry, Pathology and Agriculture. The advanced character once identified may be incorporated into a crop plant. Such studies would unveil the tremendous scope of Taxonomy and thus make it a dynamic branch of science.

REFERENCES

Bergstrom, G. (1978): In *Biological Aspects of Plant and Animal Convolution* (J. B. Harborne ed.) Academic Press, London. pp. 207-232.

Chou, C. and Y. Kuo. (1985): *J. Chem. Ecol.* **12**(6):1431-1448.

Daniel, M. and S.D. Sabnis. (1992): *Palaeobotanist* **41**:187-191.

Greca, M.D. *et al.* (1990): *Phytochemistry* **29**(6):1797-1798.

Harborne, J.B. and B.L. Turner (1984): *Plant Chemosystematics*, Academic Press, London.

Ingham, J.L. and J. B. Harborne (1976): *Nature* **260**:241-243.

Kelsey, R.G. and L.J. Locken (1987): *J. Chem. Ecol.* **13**(1):19-33.

Maisuryan, N.A. (1956): *Chem. Astr.* **50**:10863.

Matsushima, H.; Fukuui, H. and K. Arima (1973): *Agric. Biol. Chem.* **27**:1865-1870.

Nitao, J.K. (1988): *J. Chem. Ecol.* **14**(6):1515-1522.

Perez, F.J. (1990): *Phytochemistry* **29**(3):773-776.

Thein, L.B.; Heimermann, W. and R.T. Hollman (1975): *Taxon* **75**:567-568.

Yoshikawa, M. (1978): *Nature* **275**:546-547.

The Lilies - Floral Morphology, Taxonomy and Phylogeny

R.M. Pai

Department of Botany, Dr. B.A.M. University, Aurangabad, 431004.

INTRODUCTION

For well over two decades, studies on the floral morphology and anatomy of the Liliflorae are under way in this laboratory. Out of the 28 of Hutchinson's (1973) tribes of the family Liliaceae, genera of over half of them have been examined. Although this is by no means a comprehensive sample, it has helped in a board assessment of the morphological and evolutionary aspects of this admittedly primitive alliance of petaloid monocotyledons. A brief resume of the salient features of interest and importance is attemped here.

Description and Discussion

Perianth: The perianth is of six members usually differentiated into two whorls of three each. They are generally free, but may be united into a short as in *Hemerocallis, Smilacina, Mainanthemum* etc., or a long tube as in *Polygonatum* (Vaikos *et al*., 1981, 1989). The number of members may be rarely fewer as well as in *Mainanthemum* where in it is four.

Morphologically and anatomically, all the six members are similar and often referred to as tepals. They may be one-traced (Kniphofieae, Convallarieae, certain genera of Polygonatae, Scilleae etc.) or three-traced (Hemerocallideae, Tricyrtideae, Aloineae, Dianelleae, some genera of Scilleae, Polygonatae etc.). In the latter case, the lateral traces of individual tepals may arise separately from the axial stele or those of adjacent tepals may be united into the common commissurals (Vaikos and Pai, 1985, 1990; Vaikos *et al.*, 1978, 1981, 1985).

Comparative observations decisively demonstrate that the one-traced supply is basic in the lilies from which the three-traced supply is derived (cf. Pai, 1992). One may construe the other way to indicate reduction which does not appear plausible. It must be emphasized that in the higher taxa of the petaloid monocotyledons, e.g., the Amaryllidaceae, the Iridaceae and even the Orchidaceae, the tepals are three-traced (Pai, 1992 : Pande and Singh, 1979: Singh, 1972; Swamy, 1948).

It may also be noted that there is a discernible trend towards increase in the number of traces to the tepals from one to three as is occasionally evidenced in *Smilacina* and often in *Peliosanthes*.

The vasculature to the tepals has its utilitarian value in taxonomic delineations, e.g., *Eucomia*, *Scilla* of Scilleae, Polygonatum, *Smilacina* of Polygonatae and *Asphodelus* of Asphodeleae, have one-traced tepals. while *Dipcadi, Ornithogalum* of *Scilleae, Disporum* of Polygonatae and *Chlorophytum* of Asphodeleae have three-traced tepals. At the tribal level one may employ this feature thus: one-traced tepals in Convallarieae, Kniphofieae and three-traced tepals in Aloineae, Hemerocallideae, Dianelleae, Tulipeae etc.

Androecium: The stamens are six and usually arranged in two whorls of three each. They are generally free, may be rarely connate and sometimes adnate to the tepals as in genera of the Hemerocallideae, Polygonatae, Convallarieae, Scilleae etc. (Vaikos *et al.*, 1981, 1985, 1989a, b) or to the ovary as in the Aloineae (Vaikos and Pai, 1978). Within the genera of the lilies, a further differentiation of the two whorls into stamens of unequal length is noted. The stamens of the two whorls may also be inserted at two distinct lengths.

In the tribes Hemerocallideae, Kniphofieae, Aloineae for example, the stamens of the outer whorl are shorter. In *Disporum, Scoliopus* and *Johnsonia*, the inner stamens are shorter. This is considered as a trend towards reduction of either whorl which is more expressed in the higher taxa of the Liliflorae, e.g., the Iridaceae with inner stamens reduced and the Burmanniaceae with the outer stamens arrested in development (cf. Pai, 1992).

The anthers are extrorse as in *Gloriosa, Tricyrtis, Scoliopus, Disporum* etc. or introrse as in many of the genera studied. This is an important parameter in evolutionary considerations as will be seen later.

The staminal strand may sometimes bifurcate within the filament or within the connective as in *Hemerocallis, Tricyrtis, Gloriosa, Lilium* etc. (Vaikos and Pai, 1986a, b; Vaikos *et al.*, 1981). This is sometimes taken to indicate that the stamen is a product of fusion of two microsporophylls (Sterling, 1978) or that the stamen is derived from a dichotomously dividing branch system with terminal sporangia (Joshi, 1939). It must be emphasized that such divisions of staminal strands occur sporadically in many genera of monocotyledons. One should not read too much in this division to arrive at rather adventurous and far-fetched deductions. The interpretation that the tepals are staminal in origin (Leinfellner, 1962) is because both are one-traced is another similar inference.

Gynoecium: The gynoecium is typically tricarpellary and trilocular, e.g., Aloineae, or trilocular at the base unilocular upwards e.g., *Gloriosa* or unilocular from the very base as in *Tricyrtis* (Vaikos and Pai, 1986a : Vaikos *et al.*, 1978). The ovules are many to few per carpel and are generally arranged in two rows in a locule, or rarely in four as in *Aloe* and *Haworthia* (Vaikos, *et al.*, 1978).

The placentation is axile as in Hemerocallideae and Aloineae. It may be anatomically and morphologically parietal as in *Gloriosa, Tricyrtis*, Dianelleae, Convallarieae (Vaikos and Pai, 1985, 1986a; Vaikos *et al.*, 1978, 1981). The carpellary ventrals of adjacent carpels bear traces to the ovules of adjacent carpels (cf. Puri, 1952). Parietal placentation is characteristic of many genera of the Tulipeae (Berg, 1960, Eames, 1931, 1951; Vaikos and Pai, 1986b; Vasil and Johri, 1964). Genera of the Polygonatae show gradation from axile to parietal (Bjornstad, 1970; Vaikos and Pai, 1989).

The carpels are three-traced. They may be five-traced especially in genera of the Tulipeae (Eames, 1931, 1951, Joshi, 1950, Vaikos and Pai, 1986b; Vasil and Johri, 1964). The lateral traces may merge to form composite commissurals opposite to the septa in many genera of the tribe. They may not be united as in *Ornithogalum* of the Scilleae (Vaikos *et al.*, 1985).

The ovary is mostly superior. The outer floral whorls may be adnate for varying length to the ovary to develop a partially or wholly inferior ovary. In Aloineae and Hemerocallideae, adnation is for a short length. In *Ophiopogon* and Peliosanthes, the ovary is inferior. The traces to the different floral whorls show no fusion and the inferior ovary is appendicular. That the trend to develop an inferior ovary exists in the lilies is an important feature from the evolutionary standpoint as the higher taxa of the Liliflorae are with an inferior ovary.

A feature worthy of note and possibly of considerable evolutionary significance is the adnation of the stamens with the ovary although for a short length in genera of the Aloineae (cf. Vaikos *et al.*, 1978). As a variation, one or more stamens may be adnate to the whole of the length of the ovary and the style recalling the condition of the Apostasiaceae. This is regarded as an incipient trend towards the development of a gynostemium which is innate within genera of the lilies and more expressed and well developed in the apostasias and the orchids (cf. Pai 1992; Vaikos *et al.*, 1978) to infer a direct derivation of the orchids from the lilies.

The stigma in the lilies is generally carinal, sometimes commissural as in genera *Lloydia, Lilium* of the Tulipeae (Johri, 1966; Vaikos and Pai, 1986b). The occurrence of a commisssural stigma is rather erratic and its significance, if any, is obscure.

Floral nectaries: The liliaceous flower develops varied types of floral nectaries, e.g., tepaline, staminal and carpellary. Rarely, the flower does not have nectaries at all e.g., *Maianthemum, Convallaria, Chlorophytum tuberosum*.

Tepaline nectaries occur in *Gloriosa, Tricyrtis, Disporum, Calochortus*, the Tulipeae etc. Usually the bases of the tepals are nectariferous. In *Gloriosa* all the six tepals are nectariferous; in Tricyrtis and Calochortus the outer or inner tepals alone are nectariferous repectively. Further studies should adduce more data in this regard.

The staminal nectaries are of rather rare occurrence and are known in *Dianella*. The filament at its top and at the base of the anther is thickened and is nectariferous.

By far the most common type of floral nectary in the lilies is the carpellary or ovarian nectary. It is septal nectary in a great majority of the genera. It represents a glandular pocket within the septa of the ovary where the lateral walls of adjacent carpellary walls incompletely concrete and the lining layer is secretory. However, one or more layers of the cells lying within the lining layer or the nectary may also be associated with nectar secretion and are generally referred to as nectar parenchyma (Daumann, 1970). The number of layers of the nectar parenchyma may also be an indicator of the extent of development of the nectaries. The sepal nectary is generally secretory in its basal part, while its upper part is merely a canal through which the nectar oozes out.

The sepal nectaries may be developed from the very base of the ovary (*Hemerocallis*) or in its basal part which is the more general feature or at its top (*Theropogon*). They open in the basal part of the ovary (*Hemerocallis*) or at the top of the ovary beneath the base of the style which is the more general condition.

Sometimes, the septal glands may not just be pockets within the septa. They may develop canal-like passageways which indicate a higher degree of development and organization as is the case in *Hemerocallis*. In higher taxa, these passageways may be more pronounced in development as in *Phormium* of the Agavaceae.

Furthermore, the nectaries may be developed as just grooves in the ovary wall on the septal radii as in *Tofieldia calyculata* (McLean and Ivimey-Cook, 1964), *Zygadenus* (Daumann, 1970), *Smilacina, Theropogon* etc. (Vaikos *et al.*, 1989a, b). The external surface of the ovary may be nectariferous as in *Tofieldia palustris* (Eames, 1961). It must be emphasized that the situation in Tofieldia or Smilacina indicative of a lesser degree of specialization of the ovarian nectaries.

Association of the vascular tissue with the nectaries is a fact to be reckoned with. Whatever the type of nectary-tepaline, staminal or septal - the vascular tissue of the repective floral organs extends close to the nectaries or bears branches towards them. Sometimes, the vascular bundles the carpellary ventrals, in the case of septal glands, are divided into numerous tiny strands that extends close to the nectary or in the canal-like passageways as in *Hemerocallis* (Vaikos *et al.*, 1981).

Fahn (1953) is of the opinion that there is an acrocentripetal sequence in development of floral nectaries from the outer floral whorls to the gynoecium. Dahlgren and Rasmussen (1983) hold that the tepaline nectaries develop on account of loss of septal glands. Stebbins (1950), considers the tepaline nectaries of *Calochortus* as a case of a relatively late development with little to do with basic evolutionary trends. Eames (1961) regards the septal glands as highly specialized structures. It must be conceded that the septal nectaries are a class by themselves and restricted to the monocotyledons.

The development of floral nectaries-tepaline, staminal or carpellary - is essentially in relation to the floral biology of the plants and may not always serve as a pertinent parameter to indicate evolutionarily sequential development from the outer floral whorls to the gynoecium. The occurrence of tepaline (*Disporum*) and septal nectaries (*Polygonatum, Smilacina*) in different genera of a tribe Polygonatae and the absence of any type of floral nectary in *Maianthemum* of the same tribe, may be cited in this connection. The absence of nectaries in *Convallaria* and the presence of ovarian nectaries in *Theropogon* of the septal glands in many species of *Chlorophytum* and the absence of any type of floral nectary in *C. tuberosum* may also be seen in the same light. The pollination vector may visit the flower for nectar or as a pollen forager.

General: The above resume would demonstrate that the lilies possess a host of primitive characteristics of phyletic significance. It appears most plausible to derive the amaryllids and the irids along two independent lines of development rather than in a linear lineage. The amaryllids evolved along a line involving the development of an invólucrate centrifugal umbel (which can be derived from the type of inflorescence obtained in *Hemerocallis* of the lilies, cf. Rickett, 1944), differentiation of the androecium into two whorls of introrse stamens, and the attainment of an inferior ovary. The major evolutionary trend is the development of the umbellate inflorescence with reduction in number of flowers rather than reduction of stamens. The lilies have a growth form of a rhizome, bulb or corm. The amaryllids exploit the bulb as the principal growth form although other types are occasionally noted.

In the second line, the development of a definite type of inflorescence is not involved so that the irids have it of varied types. But this line involves reduction of inner whorl of the stamens so that the irid flower has the three outer stamens. In addition, the anthers are extrorse. Although the ovary is mostly inferior, it is superior in *Isophysis* to testify further a separate origin for the irids. The irids exploit the cormous growth form, although other types do occur.

The occurrence of the unique amaryllidoid alkaloids also attest a separate origin for the amaryllids (cf. Pai, 1992). Although the amaryllids are submerged within the lilies in one contemporary taxonomic treatment (Cronquist, 1988), they merit a separate entity (cf. Pai, 1992).

However, when one climbs down to taxonomic considerations or affinities amongst the tribes of the lilies, the situation presents a real task. Attempts by Dahlgren *et al.*, (1985) and Takhtajan (1980) in splitting the lilies into many smaller, apparently coherent and homogeneous entities do not appear to have made much headway. Expectedly so, for a group which is basal with so many primitive characteristics assorted out in different genera, taxonomic delineation into closely - knit categories is a difficult proposition. It is only the Aloineae that is possibly the most homogeneous amongst Hutchinson's tribes that its elevation to the level of a family may be justified (Cronquist, 1988).

It may be noted that further extensive and intensive studies are in order for a better appreciation of taxonomic delineations. In the interim, one is more at home with the lumper approach to treat the lilies as broadly as possible.

REFERENCES

Berg, R.Y. (1960): *Nytt. Mag. Bot.*, **8**:189.

Bjornstad, I.N. (1970): *Nytt. Mag. Bot.*, **17**:169.

Cronquist, A. (1988): *The evolution and classification of flowering plants*, NYBG, New York.

Dahlgren, R.M.T. and F.N. Rasmussen (1983): *Evol. Biol.*, 16:255.

Dahlgren, R.M.T.; H.T. Clifford and P.F. Yeo (1985): *The families of the monocotyledons ; structure, evolution and taxonomy* Springer - Verlag, N.Y.

Daumann, E. (1970): *Feddes Repert*, **80**:463.

Eames, A.J. (1931): *Amer. J. Bot.*, **18**:147.

Eames, A.J. (1951): *New Phytol.*, **50**:17.

Eames, A.J. (1961): *Morphology of the angiosperms* McGraw- Hill, N.Y.

Fahn, A. (1953): *Phytomorphology*, **3**:424.

Hutchinson, J. (1973): *The families of flowering plants*, Clarendon Press, Oxford.

Johri, M.M. (1966): *Phytomorphology*, **16**:92.

Joshi, A.C. (1939): *Nature*, **143**:437.

Leinfellner, W. (1962): *Ost. Bot. Z.* **109**:395.

McLean, R.C. and W.R. Ivimey-Cook (1964): *Text Book of theoretical Botany, II* Longmans, London.

Pai, R.M. (1992): *Proc. 79th Indian Sci. Congr., Part II*: 1.

Pande, V.C and V. Singh (1979): *Phytomorphology*, **29**:122.

Puri, V. (1952): *Bot. Rev.*, **18**:603.

Rickett, H.W. (1944): *Bot. Rev.*, **10**:187.

Singh, V. (1972): *Can. J. Bot.*, **50**:1555.

Stebbins (1950): *Variation and evolution*. Oxford.

Sterling, C. (1972): *J. Linn. Soc. Bot.*, **65**:163.

Swamy, B.G.L. (1948): *Bot. Mus. Leaft.*, Harvard Univ., **13**:61.

Thakhtajan, A. (1980): *Bot. Rev.*, 46:225.

Vaikos, N.P and R.M. Pai (1982): *Proc. Indian Acad. Sci.*, **91**:54.

Vaikos, N.P.and R.M. Pai (1985): *Jour. Pl. Sci. Res.*, **1**:54.

Vaikos, N.P and R.M. Pai (1986a): *Indian Bot. Reptr.*, **5**:32.

Vaikos, N.P. and R.M. Pai (1986b): *J. Indian Bot. Soc.*, **65**:516.

Vaikos, N.P and R.M. Pai (1990): *J. Indian Bot. Soc.*, **13**:51.

Vaikos, N.P.; S.K. Markandeya and R.M. Pai (1981): *J. Indian Bot. Soc.*, **60**:222.

Vaikos, N.P.; S.K. Markandeya and R.M. Pai (1985): *Trends in Pl. Res.* p. 149.

Vaikos, N.P.; S.K. Markandya and R.M. Pai (1989a): *Proc. Indian Acad. Sci.*, **99**:91.

Vaikos, N.P.; S.K. Markandya and R.M. Pai (1989b): In *Plant Science Research in India* p. 697.

Vasil, I.K. and M.M. Johri (1964): *Phytomorphology*, **14**:352.

Fruit Anatomy in Relation to Taxonomy and Phylogeny of Palms

A.R. Kulkarni & R.M. Mulla

Department of Life Sciences, University of Bombay, Vidyanagari, Bombay-400 098

INTRODUCTION

The classification of family Palmae as envisaged by different taxonomists based mainly on exmorphic features is given in Table No. 1. Of the 15 major groups of palms recognised by Moore (1973) and as tribes by Uhl and Dransfield (1985), by now fruit anatomical information on ten groups is available (Essig, 1977; Essig and Young, 1979; Guerin, 1949; Kulkarni and Pandey, 1978; Murray, 1973; Padmanabhan and Regupathy, 1981; Pandey, 1979; Reddy, 1982; Reddy and Kulkarni, 1982, 1985 a,b; Uhl and Dransfield, 1985).

Observations and discussions

(A) ANATOMY

A typical palm fruit is a drupe; berried fruits are found in members of some groups. The pericarp is differentiated into an outer epicarp, the middle mesocarp and the inner endocarp. The epicarp is constituted by the epidermis and the hypodermis, the latter being few to many layered zone of collenchymatous or sclerotic cells often impregnated with tanniniferous material and sometimes interrupted by fibre strands. The mesocarp is pulpy and or fibrous. Distributed in this ground tissue are found the mechanical elements, sclereids, fibre strands, and the protective elements in the form of raphide sacs and the tannin cells/sacs. The endocarp is represented either by

Table No.1
Comparative account of classification for palmae

1	2	3	4	5	6	7	8	9	10	11	12
Uhl and Dransfield (1985)	Mcore (1973)	Potztal (1964)	Satake (1962)	Beccari and Pitchiser-molli (1956)	Saakov (1954)	Burret (1953)	Cook (1913)	Bentham and Hooker (1889)	Drude (1889)	Martinus (1850)	Tomlinson (1961)
as sub-families	as groups	as sub-families	as sub-families	as sub-families	as groups	as sub-families	as families	as Tribes	as sub-families	as sub-famiiies	as groups
I Coryphoi-deae Tribe Corypheae	Coryphoid	5 Coryphoi-deae	2 Coryphoi-deae	5 Coryphoi-deae	Corypheae	5 Coryphoi-deae Livistoninae Subgroup Sabalinae	Sabalaceae	3 Corypheae	Coryphinae Tribe Sabaleae	4 Coryhinae Subgroup Sabalinae	Sabaloid
II Coryphoi-deae	Phoeni-coid	6 Phoeni-coideae	5 Phoeni-coideae	6 Phoenicoi-deae	Phoenicoi-deae	6 Phoenica-ceae	Phoeni-ceae	2 Coryphinae Phoeniceae	Coryphinae Phoeniceae	Phoenicoid subgroup Phoenicinae	
III Coryphoideae	Borassoi-deae	3 Borassoi-deae	1 Borassoi-deae	4 Borassoi-deae	Borassoi-deae	3 Borassoi-deae	Borassa-ceae	5 Borasseae	2 Borassinae (including *Phlido-corpus*	3 Borassinae	Borassoid
IV Calamoideae	Lepido-caryoid	4 Lepido-caryoideae	3 Lepido-caryoideae	3 Lepido-caryoideae	Lepido-caryoideae subgroup Calameae	4 Lepido-caryaceae	Lepido-caryeae	4 Lepido-caryinae	3 Lepido-caryinae Tribe Mauritieae Metroxyleae Calameae	2 Lepido-caryoideae subgroups Pinnati-frondiae Flabelli-frondiae	Lepido-caryoideae
V Nypoideae	Nypoid	2 Nypoideae	10 Nypoideae	7 Nypoideae	Areceae	2 Nypoideae (only *Nypa*	Phytale-phantaceae	Areceae Genus of Dubious affinity	Phytele-phantinae	6 Genera *Heteroclita*	Nypoid

1	2	3	4	5	6	7	8	9	10	11	12
VI Arecoideae tribe Caryoteae	- Caryotoid	8 Caryotoi-deae	7 Caryotoid	8 Arecoideae subgroup Caryoteae Excluding *Orania*	Areceae	subfamily Arecoideae Tribe Caryoteae Excluding *Orania*	Caryotaceae	Areceae Tribe Caryotideae Including *Orania* & *Scleros-perma*	Ceroxylinae Tribe Arecineae Subtribe Caryoteae Including *Orania*	Arecinae subgroup Caryotinae	Caryotoid Excluding *Orania*
VII Ceroxyloi-deae Tribe Cyclospa-theae	Pseudoph-oenicoid	Arecoi-deae Tribe Ceroxy-leae	Arecoi-deae Tribe Ceroxy-leae	Arecoi-deae	Areceae	Miscellan-eous genera in subfamily Arecoideae	Pseudoph-oenicaceae	- -	Ceroxylinae Tribe Arecineae Subtribe-Morenieae		Genera Incertecedis
VIII Ceroxyloi-deae Tribe Ceroxyleae	Ceroxyloid	Arecoideae Tribe Ceroxy-leae	Arecoideae Tribe Ceroxy-leae	Arecoideae	Areceae	subfamily Arecoideae Tribes Kentieae Ptychospermae	Ceroxy-laceae	Areceae Subtribe Ceroxy-leae	Ceroxylinae Tribes Arecineae		*Ceroxylon* in genera Incertecedis
IX Ceroxyloi-deae Tribe Hyphorbeae	Chamae-doreoid	Arecoi-deae Tribes Ceroxy-leae and Chamae-doreae	Arecoi-deae Tribes Ceroxy-leae and Chamae-doreae	Arecoi-deae Subgroup Clinostig-meae	Areceae	Subfamily Chamae-doreae	Chamae-doraceae Synaecan-thaceae	Areceae Subtribe Chamae doreae	Ceroxylinae Tribe Arecineae Subtribe Morenieae	(1) Aracinae	Chamae-doroid
X Arecoideae Tribe Iriarteae	Iriarteoid	Arecoideae Tribe Iriarteae	Arecoideae Tribe Iriarteae	Arecoideae	Areceae	Family Arecoideae Tribe Iriarteae	Iriartaceae	Areceae Subtribes Iriarteae Wettinieae	Ceroxylinae Tribe Arecineae Subtribe Iriarteae		Iriartoid

1	2	3	4	5	6	7	8	9	10	11	12
XI Arecoideae Tribe Podococceae	Podococoid	Arecoideae Tribe Dypsideae	Arecoideae Tribe Dypsideae	Arecoideae Subgroup Dypsideae	Areceae	Subfamily Arecoideae Tribe Kentieae(5)	Geonomaceae	Areceae Subtribe Geonomeae	Ceroxylinae Tribe Arecineae Subtribe Geonomeae		Arecoid
XII Arecoideae Tribe. Areceae	Arecoid	7 Arecoideae	6 Arecoideae	Arecoideae Subgroups Areceae Clinostigmeae Dipsideae Iguanareae Kentieae Linospadiceaae Ptytospemae	Areceae	Subfamily Arecoideae Tribes Kentieae and Ptychospermae	Melotriaceae Arecaceae Manocaviaceae Acristaceae	Areceae Subtribes Clinostigmeae Eguanureae Linospadiceae, Melotrireaee Oncospermae Ptychospermae	4 Ceroxylineae Tribe Arecineae Subtribe Areceae		Arecoid
XIII Arecoideae Tribe Cocoeae	Cocosoid	Cocosoideae	9 Cocosci-deae	2 Cocoideae	Areceae Subgroups: Arecastrinae Jubaeinae Butiinae	Cocoidae Tribes: Attaleae Bacterideae Cocoideae	Cocoaceae	Cocoineàe Subtribes: Eucocoineae.Elaeideae Bactrideae	Ceroxylinae Tribe Cocoineae Subtribe Bacterideae Elaeideae Attaleae	5 Cocoinae Subgroups: Genera *Aculeata* ı Genera Inermia	Bactroid Cocoid Elaeoid
XIV Arecoideae Tribe Geonomeae	Geonomoid	Arecoideae Tribe Geonomeae	Arecoideae Tribe Geonomeae	Areceae Arecoideae Subgroup Oranieae	Subfamily Arecacoidae	Arecoideae Tribe Geonomeae	8 Geonomaceae	Areceae Subtribe Geonomeae	Ceroxyline Tribe Arecineae	Arecinae *Alveolores*	Arecoid
XV Phytelephantoideae	Phytelephantoideae	9 Phytelephantoideae	8 Phytelephantoideae	8 Phytelephantoideae	Phytelephantoi deae	Phytelephas included in Arecoideae	Phytelephantaceae	Areceae genera of Dubious Affinity *Monicaria* *Leopoldinia*	5 Phytelephantinae	Genera *Heteroclita*	Phytelephantoid

slightly or highly differentiated locular epidermis or a cylinder of crystalliferous or noncrystalliferous sclerotic strands or sclerosed parenchyma mixed with sclerosed shealths of fibre and fibrovascular strands rarely involving the outer seed coat. The seed coat is smooth or variously ridged and furrowed or grooved. It is either single or double zoned, vasculated or nonvasculated. The endosperm is either homogenous or ruminated, the latter may be due to infolding or ingrowth of the seedcoat into the endosperm. Chalazal tanniniferous branched or unbranched ingrowths are common in the endosperm.

(B) FRUIT ANATOMICAL FEATURES OF MAJOR PALM GROUPS:

Coryphoid Palms

Epidermis, in surface view, is mostly differentiated into polygonal thick-walled areas (fig. I, 1) Sclereids organised into a continuous or discontinuous subhypodermal cylinder (fig.I,1), and/or scattered in the tissue of the mesocarp; coloured cells and raphide sacs present (fig.I,3), fibre bundles absent; endocarp single-layered, represented by locular epidermis alone, or many-layered, consisting of crystalliferous or non-crystalliferous sclerotic cells alone (fig. I,2 *endp*); vascular strands with poorly differentiated fibrous sheath, situated exterior to the endocarp or interior to the endocarp (fig.I,2 *vb*); seed coat single zoned, nonvasculated; chalazal tanniniferous ingrowth present in majority (fig. I,3 *cr.*); endosperm mostly homogenous, solid.

Phoenicoid palms

Epidermis in surface view not differentiated into polygonal thickwalled areas; a continuous or discontinuous single layered subhypodermal cylinder of sclereids (fig. I,4 *sc*); a continuous or discontinuous many-layered tannin cylinder in the middle part of mesocarp

EXPLANATION OF FIGURES

FRUIT ANATOMY OF PALMS

Fig.I

1,2 & 5 *Cocothrinax argentia*.

1. Epidermis in surface view x 400.
2. Fruit wall secstor in T.S. x 100.
3. T.S. of fruit gross x S.

3. *Licuala peltata*. T.S. of fruit and seed in sector x 10.

4 & 8. *Phoenix rupicola* T.S. of fruit and seed in sector x 50. Passing through the groove.

6. *Borassus flabellifer* part of a endocarp in T.S. & 200.

7. *Latania commersonii*, T.S. of the fruit in sector x 20.

9. *Caryota mitis*. T.S. of fruit and seed, sector x 40.

10. *Rafia ruffia* Fruit x 1.5.

11. *Calamus rotang* T.S. of pericarp in sector x 100.

12. *Nypa fruiticans* Fruit sector in T.S. x 20.

13. *Hyophorbe americaulis* T.S. of fruit in sector x 50.

14. *Areca triandra* Fruit wall in sector x 20.

15. *Ptychosperma singaporensis* T.S. of fruit and seed in sector x 20.

16. *P. elegans* T.S. of seed in sector x 160.

17. *Elaeis guinensis*, T.S. of fruit and seed in sector x 8.

(*edsp* - endosperm; *endp* - endocarp; *Fvb* - fibro-vascular bundle; *iop* - inner outgrowth of pericarp; *rs* - raphide sac; *Sc* - Sclereid cell/cylinder; *Sec* - Seed coat; *TC* - tannin cell/cylinder; *vb* - vascular strand).

Fig. I

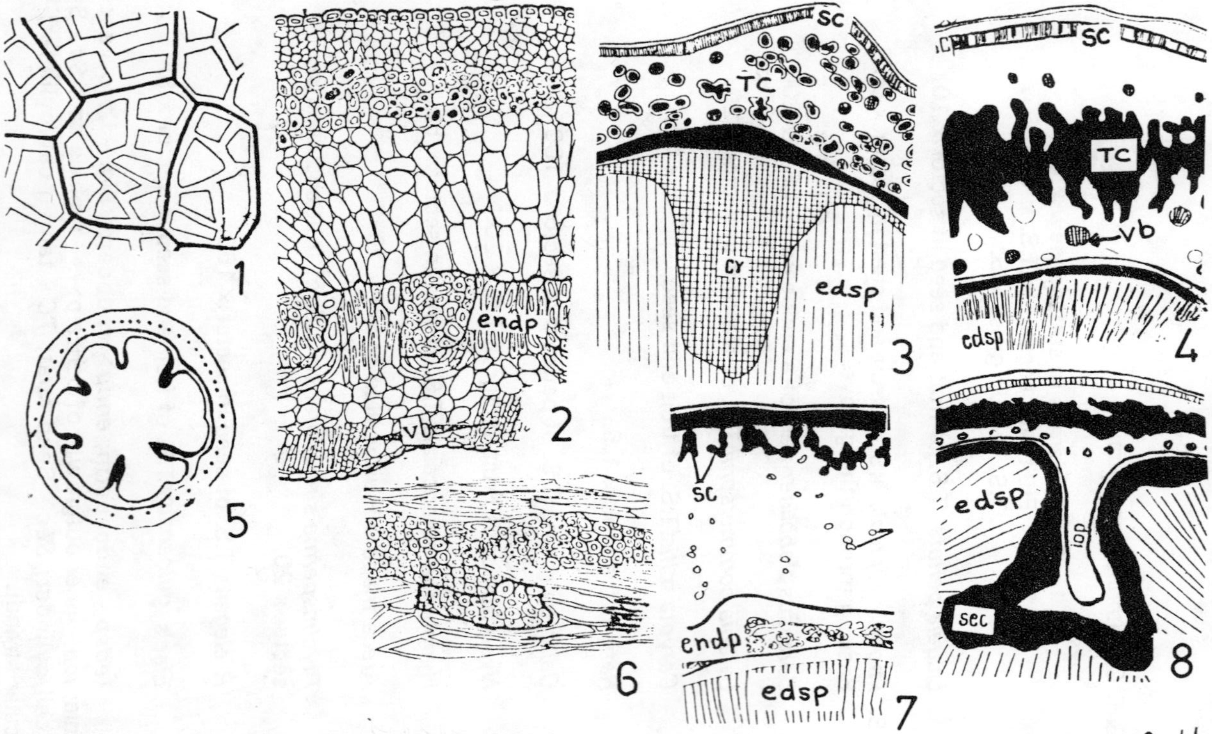

Contd----

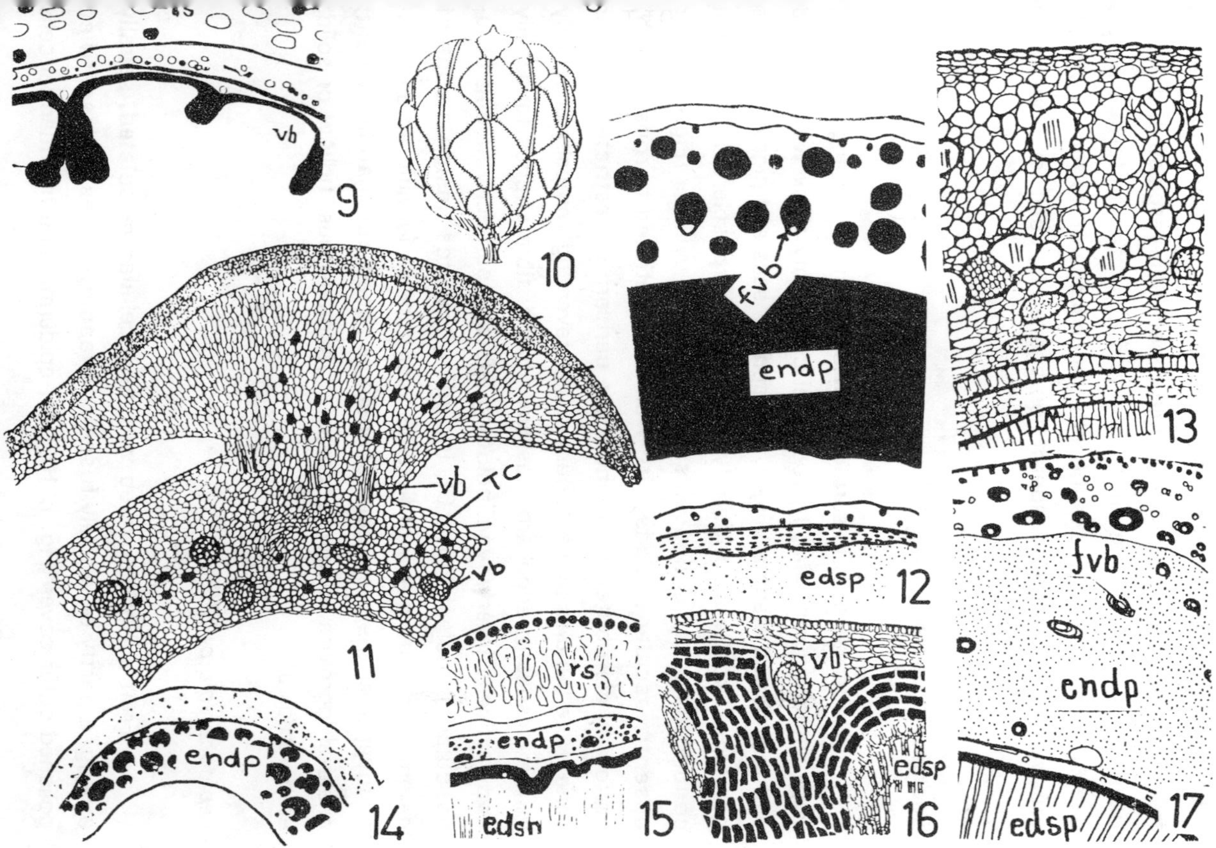

vb
9
10
fvb
endp
13
vb
TC
vb
11
edsp
12
fvb
endp
endp
rs
endp
vb
edsp
edsp
14
edsn
15
16
17

(fig. I, 4*TC*): vascular bundles with poorly differentiated fibrous sheath, in two rows, one on either side of tannin cylinder (Fig. I,4 *vb*); fibre bundles absent;endocarp constituted by the locular epidermis, separating as a thin papery layer from the mesocarp as well as from the seed coat, seed coat single zoned, non-vasculated, differentiated into outer nontanniniferous and inner tanniniferous parts, deeply infolded along the raphe into the endosperm to form a ventral groove (Fig.I,8 *iop*); endosperm homogeneous, solid, embryo in the mid dorsal part of the seed.

Boarassoid palms

Epidermis cells in surface view squarish to rectangular, filled with colouring material; hypodermal cell highly thick walled, filled with pigmented material (Fig. I, 7); raphide sacs and pigmented cells numerous, scattered uniformly in the mesocarp; sclereids forming a continuous or discontinuous sub-hypodermal cylinder (Fig.I,7 *sc*) or absent; vascular strands with well differentiated fibrous sheath, uniformly distributed throughout the pericarp or restricted to inner zone; fibre bundles present, numerous, traversing vertically or radially in the mesocarp; endocarp massive, fibrous, fibre strands tangetially and vertically interwoven including islets of parenchyma and fibrovascular bundles (Fig.I, 6), including towards inside a zone of parenchyma with a row of fibrovascular bundles forming an additional cover over the seed; seed coat single zoned, ruminations absent; chalazal ingrowth absent; endosperm homogenous, hollow or solid; embryo apical.

Lepidocaryoid palms

Fruit surface covered with multicellular vasculated, spirally arranged partly or wholely fused overlapping scales with scerified polished outer cover (Fig. I, 10, 11); epidermal and hypodermal cells of scales highly sclerosed; the fruit wall homogenous, not differentiated into meso-and endocarp; raphide sacs and sclereids few, pigmented cells common in tissues of scales; vascular bundles with poorly differentiated fibrous sheath, in a single ring (Fig. I, 11 *vb*); all the cells of the fruit wall more or less sclerosed forming a bony cover over seed; seed coat two-zoned, outer vasculated, nontanniniferous, inner non-vasculated, tanniniferous; ruminations due to ingrowth of seed coat present in *Raphia*; endosperm homogenous solid, the peripheral cells with prismatic crystals.

Caryotoid palms (Fig.I,9):

Epidermal cells, in surface view, polygonal; hypodermis narrow, collenchymatous; mesocarp with a continuous cylinder of subhypodermal sclereids; raphide sacs prominent, several large ones in mid mesocarp and 1-3 or more rows of small ones in the inner mesocarp of compactly arranged small-sized perenchyma cells abutting on endocarp; vascular bundles with poorly differentiated fibrous sheath, scattered evenly or in two rows in the mesocarp; endocarp constituted by locular epidermis, one-layered, separating as a gelatinous layer from mesocarp, cells rectangular, thick-walled, containing several silica crystals or wholly filled with solid tannin block; seed coat two-zoned, outer nontanniniferous, vasculated, inner tanniniferous nonvasculated; endosperm ruminated in *Caryota*.

Chemaedoreoid palms (Fig. I, 13):

Epidermis, in surface view, differentiated into thick-walled polygonal areas; hypodermis collenchymatous; mesocarp with 2-4 rows of large raphide sacs, sclerotic cells absent, pigmented cells absent, vascular bundles with poorly differentiated fibrous sheath in two rows, restricted to the inner part of the mesocarp, endocarp organized by locular epidermis, single-layered, forming a thin chartaceous cover over the seed, cells highly sclerosed palisaded, each with a single circular, spinulose crystal; seed coat single-zoned, non-vasculated, differentiated into outer nontanniniferous and inner tanniniferous parts; ruminations absent in *Hyophorbe*; endosperm homogenous, soiid.

Arecoid palms (Fig. I, 14, 15):

Epidermis mostly of polygonal cells in surface view, not differentiated into thick-walled polygonal areas; hypodermis narrow, collenchymatous, cells mostly squarish to rectangular, raphide sacs common in majority, coloured cells mostly present ; sclereids present in some, organised to form a subhypodermal ring, or freely scattered in middle zone; fibrovascular bundles with well developed fibrous sheaths, distributed evenly throughout the fruitwall or restricted to the inner-zone; endocarp formed of sclerosed fibrous sheaths of all or inner rows of fibrovascular bundles along with the sclerosed parenchyma in-between, extending to and including locular epidermis; locular epidermis palisaded, sclerosed and crystalliferous in many; seed coat mostly two-zoned, outer non-tanniniferous, vasculated, inner tanniniferous, ruminations in the form of ingrowths of inner seedcoat present in majority (Fig. I, 16); endosperm solid.

Cocosoid palms (Fig. I, 17):

Epidermis of rectangular longitudinally extended cells in surface view, with markedly thickened anticlinal walls, cells highly papaillate in crosssection with a sinuous line of cuticle over the outer face; hypodermis interrupted by row of fibre bundles in majority; subhypodermal parernchyma with several rows of successively larger fibre bundles; fibrovascular bundles with very well developed fibrous sheaths, almost evenly distributed throughout the fruitwall; midmesocarp typically with numerous raphide sacs, mucilage sacs or cavities, pigmented cells few; endocarp massive, consisting of sclerotic parenchyma and fibrous sheaths of vascular bundles extending upto and involving sclerotic locular epidermis and the endocarp, basal, apical or lateral, solely consisting of sclerosed palisaded locular epidermis; seed coat 2-zoned, outer non-tanniniferous, vasculated, closely adherent to the inner surface of the endocarp, inner tanniniferous, non-vasculated; ruminations rare; endosperm mostly homogenous, solid or hollow; embryo basal, apical or lateral.

C) FRUIT ANATOMY AND MAJOR GROUPS OF PALMS

In different systems of classifications, Coryphoid, Phenicoid and Borassoid palms are either treated as sub-divisions of Coryphoid line or as independent taxa. The fruit anatomical evidences reveal that while a number of features shared by phoenicoid palms are individually found in Coryphoid palms Borassoid palms present unique features ralated to endodcarp development and the structure, not shared by any other groups of palms with possible exception of *Eugeissona* and *Nypa*. Hence it is suggested that while Coryphoid and Phoenicoid groups could be treated as subgroups of one larger taxon, the Borassoid group merits treatment as an independent taxon as suggested by Drude (1987) and Martius (1850).

Lepidocaryoid palms as such, are recongnised by the presence of persisting overlapping shining scales covering fruit surface. The fruit anatomical account available is restricted to *Raphia*, *Calamus*, *Metroxylon* and *Eugeissona*. However, as stated by Uhl and Dransfield (1985) there are not much histological differentiation in fruit tissue (except *Eugeissona*); entire fruit wall getting sclerified to discharge the functions of the endocarp, making it a distinctive group of palms as suggested by all the taxonomists.

Except Drude (1887) and Martius (1823-1850), all other taxonomists have treated monotypic *Nypa* as a distinctive group of palms. In its composite structure of endocarp and numerous fibre and fibrovascular bundles of the mesocarp (Fig.I, 12) it resembles Borassoid palms rather than *Phytelephas* However, overall evidences justify it's treatment under a distinctive group of a subfamily status as indicated by Uhl and Dransfield (1985).

Caryotoid palms are considered either as a major group of palms or a subgroup of Arecoid stock. They share many features of Coryphoid palms in fruit anatomy while seed structure is similar to Arecoid palms. Their most important feature is the manner of distribution of raphide sacs in the fruit wall, especially their aggregation around the endocarp. Amongst Arecoid stock this feature is shared by Ceroxyloid and Chamaedereoid palms. On the whole, therefore, it appears better to treat them as a distinctive group of palms equivalent in rank with other major groups.

Arecoid palms have been recognised as a distinctive group of palms by all the taxonomists. A set of well-defined fruit anatomical features are shared by them. However, within the group trends of progressive specialization in the organization of the endocarp exhibited by their members (Kulkarni and Pandey, 1978) could very well form the basis to recognise the subgroups.

Cocosoid palms have either been regarded as one of the distinctive groups of palms or as a subgroup of Arecoid stock. Major mature fruit anatomical features of Cocosoid palms parallel those of Arecoid palms. 'Eyes' found in the endocarp can be campared to those of Euganurinae (Uhl and Dransfield, 1985) of Arecoid stock. However, unlike in Arecoid palms the endocarp in them originates as distinctive meristematic cylinder surrounding the locules. In majority of them it gets ontogenetically fused with the outer integument of the ovule as it develops in the seed. These features being distinctive of Cocosoid palms warrant their recognition as a distinct group. The common features between them and Arecoid palms possibly represent parallel evolution from a common ancestral stock.

D) FRUIT ANATOMY AND MINOR GROUPS OF PALMS

Apocarpous genera of Coryphoid palms have been grouped in five units by Moore (1973) while Uhl and Dransfield (1985) have kept them all under subtribe Trinacinae of tribe Coryphaeae. Amongst these, the Thrinax unit consisting of *Thrinax Cocothrinax* and *Zombia* has typically the same anatomical features, with their endocarp originating as a sclerotic cylinder far away from locular epidermis and all vascular strands restricted to the zone of fruit wall between locular epidermis and the endocarp, which gets absorbed as the fruit matures. In these respects, Thrinax unit appears to be homogenous and its genera can also be diagnosed on the fruit anatomy.

Within Coryphoid palms, *Sabal* is recognised by complete syncarpy extending to style and is kept as a separate alliance/subtribe (Moore 1973; Uhl and Dransfield, 1985). Anatomically the genus is recongised by the endocarp solely organised by the locular epidermis and the subhypodermal sclerotic cylinder.

Biradar and Mahabale (1969) have brought out significance of seed anatomy in the taxonomy of *Phoenix*.

The three genera of Caryotoid palms can at sight be distinguished in fruit anatomy using characters of distributions patterns of vascular strands in the mesocarp and the nature of the endocarp.

Drude (1889) includes genus *Orania* in his subtribe Caryoteae while rest of all the taxonomists have aligned this genus with Arecoid palms. In having pronounced development of fibrous sheaths around vascular strands and in its structure of endocarp, the genus undoubtly reveals its affinity with Arecoid palms.

Amongst Lepidocaryoid palms, *Eugeissona* stands apart in fruit anatomy in its thick woody endocarp of Borassoid type with 6-12 internal longitudinal flanges penetrating the seed and the development of conspicuous fibrous sheaths around the vascular strands (Gandhi, 1989). These justify its taxonomic distinctness within Lepidocaryoid palms as done by Moore (1973) and Uhl and Dransfield (1985).

Pseudopheonicoid, Ceroxyloid and Chamaedoreoid palms have been grouped together to form subfamily Ceroxyloideae by Uhl and Dransfield (1985). This finds support in fruit anatomy, as all the members have an endocarp organized by modified or unmodified locular epidermis, the vascular strands have poorly developed fibrous sheath and the raphide sacs dominate in the fruit wall. The fact that in several members, the vascular strands are situated just outside the endocarp racalls situation found in Arecoid palms suggesting the affinity of Ceroxyloideae to Arecoideae. The subfamily stands as one of the primitive stocks of palms along with Coryphoid palms.

(E) EVOLUTIONARY TRENDS IN THE ANATOMY OF PALM FRUITS:

Primitive versus advanced status of each character of palm fruits is summarized in Table 2.

Table No.2
Primitive *versus* advanced status of fruit characters of Palms

	Character	Primitive status	Evolved status
FIBRO			
I	Size	Small	Large
II	Surface	Smooth	Covered with sclerified scales
III	Nature	Fleshy	Fibrous
IV	Origin	Apocarpous	Syncarpous
V	Type	Uniloculaı, one-seeded	Trilocular tri-seeded
VI	Mesocarp		
	i) Sclereids	Distributed singly	Aggregated to form sub-hypodermal cylinder
		Brachysclereids	Fiber sclereids
	ii) Raphide sacs	Distributed singly	Aggregated to form zones
	iii) Tannin cells/ sacs	Distributed singly	Aggregated to form cylinder
	iv) Fibre strands	Absent	Present
	v) Vascular strands	Without lignified fibrous sheaths	With lignified fibrous sheaths
	vi) Endocarp	Nonsclerotic, noncrystalliferous, organised by locular epidermis	Sclerotic,crystalliferous locular epidermis + sclerotic tissues
TANNINIFEROUS			
		Nonfibrous, not fused with outer seed coat	Fibrous, fused with outer seed coat
	i) Seed surface	Smooth	Ridged & furrowed/ grooved
		Not covered by inner tissue of pericarp	Covered by inner tissue of pericarp
	ii) Seed coat	Singe-zoned nonvasculated	Two-zoned vasculated
	iii) Chalazal tanniniferous ingrowth	Present	Absent
	iv) Seed coat rumination	Absent	Present

Co-relation of available evidences with fruit anatomical features suggests that the primitive palm fruit is a small berry developing from one of the three carpels of tricarpellate gynoecium. The mechanical elements of such a fruit being hypodermis of somewhat sclerotic cells, sclereids and the different type of protective elements in the form of tannin cells/sacs and raphide sacs distributed in the mesocarp; the fibrous strands and conspicuous fibrous sheaths around vascular strands entirely lacking from the fruit wall. All Coryphoid, Phoenicoid and primitive palm groups of Arecoid line of evolution like Podococcoid, Pseudophoenicoid and Ceroxyloid palms share these features. Within each of these groups one could trace the specialization trends in organization of these elements like formation of subhypodermal sclerotic cylinder in most of Coryphoid, all Phoenicoid, Caryotoid palms and some Arecoid palms; a continuous tannin cylinder of the mesocarp being typical of Phoenicoid, also shared by some Coryphoid and a few Arecoid palms; distinctive type of distribution of raphide sacs, the larger in mid-mesocarp and the smaller compactly arranged in several rows in the inner mesocarp adjacent to endocarp in all Caryotoid palms, an aggregation in the inner mesocarp also is seen in several members of Chamaedoreoid and Ceroxyloid palms (Uhl and Dransfield, 1985).

The fibrous mechanical elements in the form of vascular sheaths and fibre strands are advanced features and characterize all Borassoid, Nyphoid, Arecoid and Cocosoid palms. Trends of replacement of sclereids by these are seen in several of them. Distinctive pattern of organization of these, characterizes several subgroups of Arecoid palms. The endocarp, in its most primitive form, is organized by unmodified locular epidermis alone as seen in all Phoenicoid, majority of Ceroxyloid, Chamaedoroid and some Coryphoid palms. Trends of organization of endocarp in Palmae have already been brought out by Kulkarni and Pandey (1978). The Coryphoid palms appear to be an experimental platform for organization of types of endocarps in the family. The Borassoid, Cocosoid and possibly

Nypoid endocarp are derived from a distinct meristematic cylinder organized in mid/inner part of ovary wall and can be regarded the most advanced type. Within Arecoid line of evolution one can trace progressive specialization in organization of the endocarp starting with locular epidermis, involving successively fibrous sheaths of one to many inner rows of vascular strands alongwith sclerosis of intervening parenchyma in different subgroups. Large fibrous palm fruits with massive endocarps distinctly represent an advanced condition.

The seed coat in most primitive state is non-vasculated, single zoned, distinguished into outer non-tanniniferous and inner tanniniferous layers. This type is a universal feature of all Coryphoid and Phoenicoid palms. In Borassoid and Nypoid palms, tissues of mesocarp interior to the endocarp get abutted with the seed coat forming an additional cover over the seed surface. In rest of the palm groups the seed coat is two-zoned, outer vasculated and inner tanniniferous. Chalazal tanniniferous ingrowth is typical of Coryphoid palms, while Phoenicoid palms show deep tanniniferous invagination of the seed coat. Ingrowths of inner integument at sites of vascular strands of outer, form features of several Arecoid and few Cocosoid palms. All these, alongwith distinctive infoldings of seed coat at one or several sites of the integument form advanced features seen in a few Coryphoid (fig. I, 5) (*Cocothrinax*), several Arecoid and Nypoid palms.

(F) FRUIT ANATOMY AND EVOLUTION OF PALMS:

The schematic representation of evolution of palm groups based on overall evidences is depicted in Fig.II. Coryphoid palms are undoubtedly one of the most primitive extant palm groups, along with Pseudophoenicoid, Ceroxyloid and Chamaedoreoid palms. Rest of all the palm groups must have diverged from the primitive palm stock quite early and must have independently evolved into climax groups.

REFERENCES

Drude, O. (1887): *Die Naturlichen pflanzen familien* 1, Teil **2**: 1-93, Berlin.

Essig, F.B. (1977): *Syst. Bot.* **2**:151-168.

Essig, F.B. & B.E. Young (1979): *Syst. Bot.* **4**:16-28.

Gandhi, K.S. (1989): *Contribution to Palm Fruit Anatomy*, M. Phil. Thesis, University of Bombay.

Guerin, H.P. (1949): *Annals. Sci. Nat. Bot. Serie* **11** *10*:21-69.

Kulkarni, A.R. and S.B. Pandey (1978): *Evolutionary Botany* and *Biostratigraphy* (A.K. Ghosh Commemoration Vol.1 A.K. Sharma, ed.) 185-193.

Martius, C.F.D. (1823-1950): *Historia Naturalis Palmarum* 3, Vols., Munich.

Moore, H.E. (1973): *Gentes Herbarum* **11**(2):27-141.

Murray, S.G. (1973): *Principes* **1**(3):91-102.

Padmanabhan, D. and D. Regupathy (1981): *Principes* **25**(4):172-177.

Pandey, S.B. (1979): *Studies on Palms*: *Contributions to the anatomy of fruits and seeds*. Ph.D. thesis, University of Bombay.

Reddy, G.N. (1982): *Development Anatomy of Fruits and seeds of Palms*, Ph.D. thesis, University of Bombay.

Reddy, G.N. and A.R. Kulkarni (1982): *Geophytology* **12**(2):233-244.

Reddy, G.N. and A.R. Kulkarni (1985a): *Phytomorphology* **35**:25-33.

Reddy, G.N. and A.R. Kulkarni (1985b): *Proc. Indian Acad. Sci.* **95**:153-165.

Reddy, G.N. and A.R. Kulkarni (1986a): *J. Indian Bot. Soc.* **65**:345-352.

Reddy, G.N. and A.R. Kulkarni (1986b): *Indian J. Bot.* **9**(1):34-42.

Uhl, N.W. and J. Dransfield (1985): *Genera Palmarum, A Classification of Palms*. Allen Press, Lawrence, Kansas.

Cytological Conspectus of Indian Cyperaceae

P. Cheema and S.S. Bir
Department of Botany, Punjabi University, Patiala-147002, India.

INTRODUCTION

Sedges, the members of family *Cyperaceae*, are represented in Indian flora by 22 genera and 400 species (Santapau & Henry, 1973). Their chromosomal analysis has aroused great interest because of the discovery of a variety of chromosome numbers being reported for various species particularly those of *Cyperus, Carex, Scirpus*, etc.

First of all Sharma & Bal (1956) studied the karyology of *Cyperaceae* members from Eastern India. This was followed by Sanyal & Sharma (1972) who analysed 40 species belonging to 7 genera of tribes Scirpeae and Cyperae of the same region. Rath & Patnaik (1974, 1978) studied cytology of 67 species of sedges from Orissa state of Eastern India whereas Nijalingappa (1973, 1974, 1975, 1977), Nigalingappa *et al.* (1978), Nijalingappa & Leela Bai (1990) and Subramaniam (1988) analysed members of *Fimbristylis, Ascopholis, Cyperus, Lipocarpha, Carex* etc. from South India. From Western Himalayas cytological information about Cyperaceae is available from Mehra & Sachdeva (1975a,b, 1976).

During the last one and-a-half decades, chromosomal analysis of sedges of Punjab state, NW India has been undertaken by S.S. Bir and his students (Bir *et al.* 1982, 1986, 1988 a,b, 1990, 1991, 1992a,b,c, 1993a,b,c, 1994, Cheema, 1991, Cheema *et al.*, 1992a,b, 1993a,b, Cheema & Bir, 1994, 1995, Bir & Cheema 1994).

So far, the cumulative data show that information about the chromosome numbers of 42.5% of the total Indian sedges is available. Polyploidy is of frequent occurrence and intraspecific variations are too well apparent.

The main purpose of the present communication is to assess the evolutionary patterns of Indian sedges which constitute a very significant part of our flora. The information thus generated will help in planning of future work.

COLLECTION OF CHROMOSOMAL DATA

All the available information on chromosome numbers of Cyperaceae has been documented and analyzed from Darlington & Wylie (1955), Fedrov (1969), Love & Love (1974, 1975), Kumar & Subramaniam (1987), Index to Plant Chromosome numbers (Cave, 1958, 1959, 1961; Cave & Chisski-Hommersand 1963, 1965; Ornduff 1968; Moore 1973, 1977; Goldblatt 1981, 1984, 1985), IOPB chromosome Numbers Reports' (as presented by A. Love singly or jointly) published in *Taxon* from 1965 onward and SOCGI Chromosome number Reports I- XI (edited by S.S. Bir) published in *Journal of Cytology & Genetics* from 1983-1993 and finally, the *Biological Abstracts*, from 1971 onwards.

CHROMOSOME NUMBERS

A perusal of world-wide data on chromosome numbers has indicated that these have been determined for 754 species out of a total of 4,000 (Willis, 1966) species. This comes to only 18.82 % of the total species of sedges. It is evident that chromosome numbers vary from 2n=6 in *Fimbristylis woodrowii* (Ramakrishnan & Ramachandran, 1976) and *F. umbellaris* (Rath & Patnaik, 1977) to 2n=224 in *Cyperus cyperoids* (Tejavathi, 1988). Among Indian sedges, 170 species (42.5%) out of 400 (Santapau & Henry, 1973) have so far been worked out. Information on the total range of 'n' chromosome numbers and the frequency of chromosomal taxa carrying them has been given in Table I and figure 1.

Table 1: Frequency of various chromosome numbers in the Indian sedges

Chromosome number (n)	Taxa	% age	Chromosome number (n)	Taxa	% age
3	4	0.716	41	1	0.179
5	87	15.591	42	9	1.612
7	1	0.179	43	6	1.075
8	12	2.150	44	8	1.433
9	4	0.716	45	2	0.358
10	57	10.215	46	2	0.358
11	11	1.971	47	7	1.254
12	4	0.716	48	15	2.688
13	13	2.329	49	9	1.612
14	8	1.433	50	5	0.896
15	10	1.792	51	2	0.358
16	17	3.046	52	9	1.612
17	3	0.537	53	1	0.179
18	14	2.508	54	18	3.225
19	12	2.150	55	3	0.537
20	17	3.046	56	13	2.329
21	13	2.329	57	3	0.537
22	7	1.254	58	5	0.896
23	5	0.896	59	1	0.179
24	10	1.792	60	5	0.896
25	11	1.971	62	6	1.075
26	4	0.716	63	3	0.537
27	3	1.254	64	11	1.971
28	3	0.537	66	1	0.179
29	5	0.896	67	1	0.179
30	2	0.358	68	1	0.179
31	1	0.179	76	2	0.358
32	5	0.896	80	1	0.179
34	2	0.358	86	1	0.179
35	1	0.179	98	1	0.179
36	11	1.971	100	2	0.358
37	4	0.716	104	5	0.896
38	5	0.896	108	2	0.358
39	3	0.537	110	1	0.179
40	14	2.508	112	1	0.179

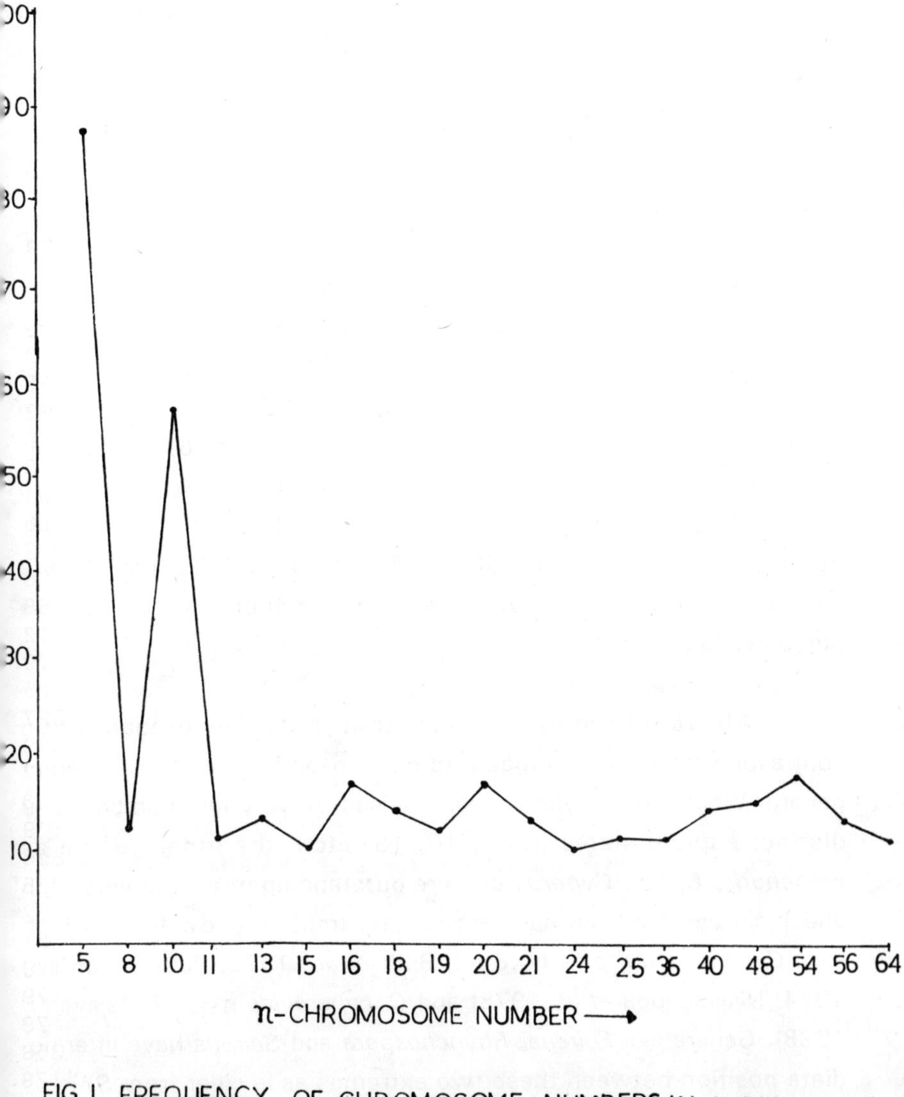

FIG. 1. FREQUENCY OF CHROMOSOME NUMBERS IN INDIAN SEDGES

A cursory look on this data points out that the most frequent 'n' number is 5 (87 taxa) followed by 10 (57 taxa). The variation of chromosome numbers within the genus and even within the species is quite common. It is supported by the fact that various genera of Cyperaceae such as *Bulbostylis* n=5, 10, 32, 36; *Carex* n=21, 23, 24, 26, 28, 29, 30, 31, 38, 42, 47, 52; *Courtoisia* n=5, 20; *Cyperus* n=5, 8, 9, 10, 11, 12, 13, 15, 16, 17, 18, 19, 20, 21, 22, 23, 24, 25, 26, 27, 28, 29, 30, 32, 34, 35, 36, 37, 38, 39, 40, 41, 42, 43, 44, 45, 46, 47, 48, 49, 50, 51, 52, 53, 54, 55, 56, 57, 58, 59, 60, 62, 63, 64, 66, 67, 68, 76, 80, 86, 98, 104, 110, 112; *Eleocharis* n=5, 7, 8, 10, 15, 23, 27, 38, 40, 100, 108; *Eriophorum* n=25, 26; *Fimbristylis* n=3, 5, 8, 10, 11, 12, 15, 16, 20, 22, 24; *Fuirena* n=18, 19, 26; *Lipocarpha* n=13, 16, 19; *Rhynchospora* n=9, 10; *Scirpus* n=5, 10, 11, 14, 15, 16, 20, 21, 22, 22, 37, 38, 39, 40, 42, 44, 47, 48, 49, 50, 51, 52, 55, 56, 57, 58, etc. show a high degree of inter and / of intraspecific variations in chromosome numbers.

Above information reveals that the family presents two contrasting features with regard to chromosome numbers in different genera. While *Fimbristylis* is characterised by very low numbers and distinct euploid series (n=5, 10, 15, etc.), the other genera as *Eleocharis, Carex, Cyperus* etc. are outstanding in having very high aneuploid chromosome numbers ranging from very low to very high e.g.; *Cyperus tenuispica* n=8 (cf. Rath *et al.*, 1973; Rath & Patnaik, 1974; Nijalingappa *et al.* 1978) and *C. cyperoides* n=112 (Tejavathi, 1988). Genera like *Fuirena, Rhynchospora* and *Scirpus* have intermediate position between these two extremes as is clear from *Scirpus* with n=5 and n=58 in *S. litoralis* (Sarkar *et al.*, 1976) and *S. tuberosus* (Bir *et al.*, 1988; 1990), respectively.

Thus like Poaceae (Gramineae), the Cyperaceae represents a group of flowering plants with highly variable chromosome numbers.

From the above account, it is concluded that the most frequent chromosome numbers within the family range between n=5 and 10. The other variations on the lower and higher side might have been there due to the phylogenetic decrease and increase in chromosome numbers at diploid level. The polyploidy and aneuploidy in the extant members coupled with this phylogenetic change explain the wide spread variation in chromosome numbers in sedges. Information on series of euploid and aneuploid numbers is provided subsequently.

CHROMOSOME SIZE

Our studies on 44 species (125 taxa) have revealed that meiotic figures of chromosomes vary greatly in size. It is clear that:

i) In four species, namely, *Cyperus bulbosus* n=46, 59; *C. compressus* n=56, 64; *C. digitatus* n=42, 52, 54 and *Scirpus roylei*, n=11, 44, there is a decrease in chromosome size with increase in chromosome number. These observations are in confirmity with observations of Walker (1938), Stebbins (1971) and Darlington (1973).

ii) Three species, namely, *Cyperus iria* n=54, 58; *Scirpus kyllingioides* n=50, 51, 52 and *S. litoralis* n=39, 42 exhibit large size of chromosomes correlated to increase in number.

iii) In the following species, increase or decrease in chromosome number has no effect on chromosome size:

Cyperus klyllingia n=48, 56, 60; *C. globosus* n=36, 40, 42, 44; *C. triceps* n=46, 48 and *Fimbristylis dichotoma* n=5, 10. Darlington (1965) stated that when there is no increase or decrease in chromosome size it seems that the multiplication of chromosomes by polyploidy is conditioned by the existence of a margin or space in the PMCs permitting an increase in the bulk of chromosomes.

iv) In the following five species chromosome size increases or decreases irregularly showing that there is no correlation between chromosome size and chromosome number:

Cyperus brevifolius n=48, 56, 62, 86; *C. alopecuroides* n=44, 48, 50, 52; *C. difformis* n=9, 17, 18; *C. rotundus* n=42, 48, 50, 52, 54, 55, 56, 58, 66, 80 and *Scirpus tuberosus* n=47, 48, 52, 56, 57, 58.

v) It is also not unusual to find that some species of a genus have large-sized chromosomes while the others possess small-sized chromosomes as is evident in *Scirpus* and *Eleocharis*. *Scirpus mucronatus* (n=22), *S. litoralis* (n= 39, 42), *S. triqueter* (n=21) and *E. palustris* (n= 8+1B) possess large chromosomes whereas *Scirpus kyllingioides* (n=50, 51, 52), *S. tuberosus* (n=47, 48, 52, 56, 57, 58) and *Eleocharis atropurpurea* (n=10) have small-sized chromosomes. Earlier, Sahni (1987) reported such variations in different species of *Aristida* and *Chloris*.

Khosla (1978) laid stress that whenever large chromosomes occur in a genus or family, which otherwise possesses small chromosomes, systematics and phylogeny of the taxon need reinvestigations. But Swanson (1957) is of the view that the differences in the diameter of chromosomes in different species of a genus indicate solely the range of genotypic change.

CYTO-/AND OR MORPHOVARIANTS

Infraspecific morphological variations with uniformity in chromosome numbers in different individuals, have been noticed by us in ten species whereas the species exhibiting cytological and/or morphological variability are sixteen in number (Table 2).

Table 2: Infraspecific chromosomal races amongst sedges of Punjab plain showing cyto-/morphotypes

S.No.	Taxon	Chromosome number
1.	*Cyperus brevifolius*	48,56,62,86
2.	*C. kyllingia*	48,56,60
3.	*C. triceps*	46,48
4.	*C. globosus*	36,40,42,44
5.	*C. alopecuroides*	44,48,50,52
6.	*C. bulbosus*	46,59
7.	*C. compressus*	56,64
8.	*C. difformis*	9,17,18
9.	*C. digitatus*	42,52,54
10.	*C. iria*	54,58
11.	*C. rotundus*	42,48,50,52,54,55,56,58,66,80
12.	*Fimbristylis dichotoma*	5,10
13.	*Scirpus kyllingioides*	50,51,52
14.	*S. tuberosus*	47,48,52,56,57,58
15.	*S. litoralis*	39,42
16.	*S. roylei*	11,44

a) **Morphotypes with the same Chromosome Numbers in Different Populations:**

Different individuals of *Cyperus kyllingia* (n = 60) *Fimbristylis bisumbellata* (n = 5), *F. ferruginea* (n = 10), *F. miliacea* (n = 5), *F. quinquangularis* (n = 5), *Scirpus mucronatus* (n = 22) and *S. triqueter* (n = 21) are with the same chromosome numbers but very often show differences in morphological characters as highlighted earlier (Bir *et al.*, 1991, 1992 b; Cheema, 1991). Two morphotypes of *C. rotundus* (n = 52) with brown and chest-nut brown spikes have the same genetic constitution (Cheema *et al.*, (1992b).

Even in case of species where the same chromosome numbers have been recorded in different populations, the individuals of the different populations show marked morphological differences although these have the same genetic constitution. Such a condition is noticed in *Cyperus globosus*, n=36, 42, *C. difformis*, n=9, 17, 18 and *C. rotundus*, n=55, 56 (Cheema, 1991; Cheema *et al.,* 1992 b, 1993b). Phenotypic differences between individuals having the same chromosome number in case of these species are well marked.

In all the above cases the morphological variations are due to variable habitat conditions as soil factor or availability of water and biotic interference. Briggs & Walter (1984) have earlier pointed out that with increase or decrease in biotic and edaphic factors, there is always a change in the morphology of plants. Such species show phenotypic plasticity in varying habitats.

b) **Morphotypes with different Chromosome Numbers**

Sixteen species taken up in the present study show infraspecific variations (Table 2). These exhibit different ploidy level and in addition, majority of the cytotypes are with aneuploid chromosome numbers.

Out of these, only three species, namely, *Fimbristylis dichotoma* (n=5, 10), *Scirpus litoralis* (n=39, 42) and *Scirpus roylei* (n=11, 44), having different cytotypes, show increase in size of vegetative and reproductive organs (Bir *et al.,* 1986, 1991) as a result of increase in chromosome number *i.e.* gigas effect is evident. Similar results were obtained in dicotyledonous plants e.g., *Emblica officinalis*, 4x and 8x (Bedi *et al.,* 1985) and *Lantana camara*, 2x and 4x (Chatha & Bir, 1986).

In twelve species the increase or decrease in plant size is irregular with the increase in polyploidy or aneuploidy. These are *Cyperus kyllingia* (n=48, 56, 62, 86), *C. brevifolius* (n=48, 56, 60),

C. globosus (n=36, 40, 42, 44), *C. alopecuroides* (n=44, 48, 50, 52), *C. bulbosus* (n=46, 59), *C. compressus* (n=56, 64), *C. difformis* (n=9, 17, 18), *C. digitatus* (n=42, 52, 54), *C. iria* (n=54, 58), *C. rotundus* (n=42, 48, 50, 52, 54, 55, 56, 58, 66, 80), *Scirpus kyllingioides* (n=50, 51, 52) and *S. tuberosus* (n=47, 48, 52, 56, 57, 58). Changes in phenotypic characters in all these species cannot be linked with genetic constitution.

One species, *C. triceps* has two distinct cytotypes with n=46 and 48. These do not show any increase or decrease in morphological parameters with increase in chromosome number.

The taxonomic treatment of such cyto-/and or morphovariants has been variously discussed. According to Stebbins (1950) and Rogers & Apan (1969) polyploid races should be treated as separate taxonomic entities since they are reproductively distinct from their diploid ancestors. Smith (Mehra, 1982) and Davies & Heywood (1963) are not in favour of regarding the cytotypes as distinct species. Mehra (1982) while discussing the grasses of eastern Himalayas, expressed the opinion that a cytotype should have distinct and specific status only when a taxon shows sharp and easily identifiable morphological characters. He opined that more correctly it should be different from its parents and then it may be regarded as subspecies or incipient species.

Evidently, the taxonomy of all such species needs to be revised and if possible the type material of each species should be identified with a particular cytotype or biotype.

BASIC (OR BASE) NUMBERS

As far as Cyperaceae is concerned, a lot of confusion exists about the basic numbers. x=5 is supposed to be the original base number of the whole family (Wulff, 1939). This is supported by Mehra & Sachdeva (1975). Haploid chromosome number n=3 reported for *Fimbristylis woodrowii* (Ramakrishnan & Ramachandran, 1976) and *F. umbellaris* (Rath & Patnaik, 1977) can be considered to be the result of aneuploid reduction (Ramakrishnan & Ramachandran, 1976).

An analysis of various basic numbers of Cyperaceae on world-wide basis reveals that many genera show dibasic, tribasic or polybasic nature (Table 3). Similarly, certain genera are characteristic in having related basic numbers as X=5, 6, 7, 8, 9, 10 etc. while others have unrelated numbers as x= 9, 19, 29. Such a great variation in basic numbers clearly brings out the role of aneuploidy and neopolyploidy in the evolution of these numbers.

In the Indian sedge genera the basic chromosome numbers vary from x=5 to x=60 and out of these x=5, 7, 8, and 9 are more common. We discuss below seven genera which have been analysed by us:

a) **Bulbostylis** Kunth

Except for *Bulbostylis capillaris*, only x=5 is suggested as basic chromosome number for the genus (Darlington & Wylie, 1955). Tanaka (1937) recorded 2n=48 for *Bulbostylis capillaris* from North America whereas Sharma (1962) reported 2n=72 for this very species from Western Himalayas. Mehra & Sachdeva (1975) opined that this species appeared to have arisen through a continued process of polyploidy and aneuploidy. Presently studied one species also confirms to x=5 as basic chromosome number.

b) **Carex** Linn.

The chromosome numbers, recorded for the largest genus of the family, range between 2n=12 to 2n=112. The occurrence of this unusually wide range of chromosome numbers which are not in multiples of any common basic number led earlier workers to postu-

Table 3: Data on basic numbers in genera of Indian *Cyperaceae*

Basic Number	Genera
X=5	*Bulbostylis, Fimbristylis**, *Eleocharis**, *Eriophorum**, *Carex, Fuirena** (6)
X=6	*Fimbristylis** (1)
X=7	*Scirpus** (1)
X=8	*Fimbristylis**, *Eleocharis**, *Cyperus**, *Scirpus**, *Rhynchospora** (5)
X=9	*Fimbristylis**, *Eleocharis**, *Eriophorum**, *Cyperus**, *Schoenus** (5)
X=11	*Fimbristylis**, *Schoenus** (2)
X=13	*Scirpus**, *Rhynchospora** (2)
X=19	*Eriophorum**, *Schoenus** (2)
X=20	*Scirpus** (1)
X=21	*Scirpus** (1)
X=23	*Lipocarpha* (1)
X=26	*Kobresia** (1)
X=27	*Schoenus** (1)
X=29	*Eriophorum**, *Fuirena**, *Kobresia** (3)
X=36	*Kobresia** (1)
X=38	*Schoenus** (1)
X=60	*Kyllinga* (1)

* Genera with more than one basic number

late different basic numbers. Heilborn (1939) considered x=7 as the basic chromosome number while Wahl (1940) suggested x=5, 6, 7 and 8. Supporting Heilborn, he believed that many of the species are derived by secondary balancing at a number below or just above the euploid numbers from species with a basic number of 7 which is probably the most common one. However, Tanaka (1948) proposed the basic numbers of 3, 4, 5, 6, 8, 9, 10 and 12 and treated x=3, 4 and 5 as primary base numbers from which others are derived secondarily through polyploidy. According to Love *et al.* (1957), all chromosome numbers within Cyperaceae in general and *Carex* in particular have been derived from the original basic number x=5 by agmatopoidy. Presently, ploidy level of various taxa has been calculated based on x=5 as the base number.

c) **Cyperus** Linn.

It is the second largest genus in the family. Darlington & Wylie (1955) suggested x=8 and 9 for *Cyperus* (*sensu stricto*). Rath & Patnaik (1974) confirmed x=8 and 9. They also recorded a new base number x=13 for the genus (*C. pulcherrimus* Willd. with n=13). Mehra & Sachdeva (1976) confirmed the existence of x=8 and 9 for the genus while Subramaniam (1988) considered x=8 as the primary base number for the family. According to him x=8 should have given rise to other basic numbers x=9, 10 and 11 by means of aneuploidy.

As far as basic numbers of different sections or subgenera recognised under genus *Cyperus* (*Pycreus, Kyllinga, Juncellus* and *Mariscus*) are concerned there is hardly any demarcation between them. The consensus of opinion is to regard x=8, 9 as basic numbers for *Cyperus* (*sensu lato*).

d) **Eleocharis** R.Br.

Darlington & Wylie (1955) suggested three basic numbers x=5, 8 and 9 whereas Love & Love (1974) supported x=5 and 8 as basic numbers. x=5 is suggested as basic number by Sanyal & Sharma (1972) and Rath & Patnaik (1978). Indian species of *Eleocharis* support x=5, 8 and 9 as base numbers for the genus.

e) **Eriophorum** Linn.

Mehra & Sachdeva (1975) suggested x=26 for the genus based on recording of n=26 in *Eriophorum comosum*. For the presently studied one species with n=26, the writers have taken x=5 as the basic number.

f) **Fimbristylis** Vahl

Tanaka (1940) reported four basic numbers viz. x=5, 6, 8 and 11. Darlington & Wylie (1955) also suggested the genus to be tetrabasic. Patnaik & Rath (1983) suggested x=3, 5, 8 and 11 but Sharma & Bal (1956) and Sharma (1962) found that all the Indian species confirm to only one base number x=5 Except for *Fimbristylis falcata* with n=11, 22 and *F. paupercula* n=22, all other species listed presently have x=3, 5 or 11 as base number.

g) **Scirpus** Linn.

Darlington & Wylie (1955) did not mention any base number for the genus. Love & Love (1975) reported x=8 for the genus. Further different basic numbers have been suggested (Table 3) as x=7, 13, 20, 21 etc. Sanyal & Sharma (1972) suggested x=7, 8 as base number but Mehra & Sachdeva (1975) assigned x=14. Patnaik & Rath (1983) state x=5 as basic number on the basis of presence of n = 15 in *S. supinus* and n = 20 in *S. squarrosus*. Three basic numbers x = 5, 7 and 8 are often in euploid series. The present authors' observations have shown that various taxa are based on x = 7 and 8.

So it is concluded that various genera of Cyperaceae are either dibasic, tribasic or polybasic. Further x = 5, 7, 8 and 9 are of frequent occurrence whereas other numbers x = 3, 6, 11, 13 etc. are rare.

POLYPLOIDY

The importance of polyploidy in evolution and speciation of plants has been stressed from time to time.

a) Incidence of Polyploidy

Sedges exhibit a wide range of chromosome numbers (from 2n = 6 to 2n = 224) which is clear indication of the prevalence of polyploidy. On world-wide basis the overall polyploidy in sedges has been estimated to be 77% (Goldblatt, 1980), 94.7% (Bir *et al.*, 1988a) and 90.4% (Cheema, 1991).

As far as Indian sedges are concerned, 89% of them are noticed to be polyploids by Bir *et al.* (1988a) whereas 85.5% are polyploids according to Cheema (1991). Present data show that 72.3% of the Indian sedges are polyploid and the frequency of polyploids is the highest in *Carex* (100%) and *Fuirena* (100%) followed by *Scirpus* (95%), *Cyperus* (89.5%), *Bulbostylis* (75%) and *Eleocharis* (50%) whereas *Fimbristylis* has only 19.4% polyploid species. Out of total Indian taxa studied uptil now, only 20.25% are diploid and a majority of them (79.74%) are polyploids.

In a few species there is no evidence of polyploidy: *Bulbostylis barbata* (n=5), *Cyperus decumbens* (n=9), *Eleocharis ovata* (n=5), *E. palustris* (n=7, 8, 8+1B), *E. retroflexa* (n=5), *Fimbristylis acicularis* (n=5), *F. acuminata* (n=5), *F. aestivalis* (n=5), *F. bisumbellata* (n=5, 8), *F. barbata* (n=5), *F. cinnamomentorum* (n=5), *F. cyperoides* (n=5), *F. dipsacea* (n=5), *F. globulosa* (n=5), *F. umbellaris* (n=3), *F. miliacea* (n=5), *F. monospicula* (n=5), *F. narayanii* (n=3), *F. schoenoides* (n=5), *F. setacea*(n=5), *F. sub-bispicata* (n=5), *F. tristachya* (n=5), *F. woodrowii* (n=3,5), *Rhynchospora corymbosa* (n=9) and *R. wightiana* (n=10).

About 22 species have both diploid as well as polyploid cytotypes. Hence these are not considered in calculating the % age of diploid or polyploid species.

Species with diploid as well as polyploid taxa are:

1.	*Courtoisia cyperoides*	n=5	n=20
2.	*Cyperus difformis*	n=9	n=17,18
3.	*C. flavidus*	n=8	n=10,38
4.	*C. inundatus*	n=5	n=29,30
5.	*C. haspans*	n=8	n=13,18
6.	*C. procerus*	n=9	n=13,32,52
7.	*C. tenuispica*	n=8	n=18
8.	*Fimbristylis cymosa*	n=5	n=16,24
9.	*F. dichotoma*	n=5	n=10,16
10.	*F. falcata*	n=11	
11.	*F. ovata*	n=5	n=10
12.	*F. polytrichoides*	n=5	n=10
13.	*F. quinquangularis*	n=5	n=10
14.	*F. ferruginea*	n=5	n=10,16
15.	*F. complanata*	n=5	n=10
16.	*F. squarrosa*	n=5	n=10,12

17.	*F. spathacea*	n=5	n=10
18.	*F. tetragona*	n=5	n=10
19.	*Eleocharis chaetaria*	n=5	n=10
20.	*Eleocharis atropurpurea*	n=5	n=10
21.	*E. geniculata*	n=5	n=10,15,40
22.	*Scirpus lateriflorus*	n=5	n=11,15,40

Table 4 reflects that polyploids are dominant in almost all the Indian genera. Various levels of ploidy show that 4x, 6x, 8x and 12-16 x grades of ploidy are very frequent and *Carex* (18x, 20x), *Cyperus* (18x, 20x, 22x, 26x, 28x) and *Eleocharis* (18x, 20x, 24x) show still higher grades of polyploids.

b) **Infraspecific Polyploidy:**

As many as 50 species of Indian Cyperaceae have been found to show the phenomenon of infraspecific polyploidy. These are *Cyperus alopecuroides* 6x, 12x, 14x, based on x=8; *C. alulatus* 10x, 14x based on x=8; *C. bulbosus* 10x, 12x, 14x based on x=8; *C. castaneus* 6x, 12x based on x=8; *C. compactus* 4x, 6x, 10x, 14x based on x=8; *C. compressus* 12x, 14x, 16x based on x=8; *C. corymbosus* 14x, 16x based on x=8; *C. cyperoides* 10x, 16x, 28x based on x=8; *C. difformis* 2x, 4x based on x=8; *C. digitatus* 10x 12x based on x=9; *C. eleusinoides* 8x, 12x based on x=8; *C. esculentus* 14x, 26x based on x=9, *C. flavidus* 2x, 4x, 8x based on x=8; *C. globosus* 8x, 10x based on x=8; *C. fuscus* 6x, 8x based on x=8; *C. haspans* 2x, 4x based on x=8; *C. iria* 8x, 12x, 14x based on x=9; *C. kyllingia* 4x, 10x, 12x, 14x, 16x based on x=8; *C. laevigatus* 10x, 12x based on x=9; *C. nutans* 6x, 8x, 12x based on x=9; *C. paniceus* 10x, 14x, 22x based on x=9; *C. pilosus* 4x, 8x based on x=9; *C. polystachyus* 12x, 14x based on x=8; *C. procerus* 2x, 4x, 8x, 12x based on x=9; *C. nitens* 6x, 10x based on x=9; *C. rotundus* 4x, 10x, 12x, 14x, 16x, 18x, 20x based on x=8

Table 4: Cytological information about the Indian sedges

S.No.	Genera*	Ploidy level	Known n-chromosome numbers (figures in parenthesis give number of taxa)
1	2	3	4
1.	*Ascopholis* (a:1,b:1,c:0,d:1,e:1)		40(1)
2.	*Bulbostylis* X=5 (a:4,b:4,c:1,d:3,e:6)	2X,4X,12X,14X	5(3),10(1),32(1),36(1)
3.	*Carex* X=5 (a:148,b:14,c:0,d:14,e:16)	8X,10X,12X,16X, 18X,20X	21(2),23(2),24(3),26(1),28(1),29(1),30(1), 31(1),38(1),42(1),47(1),52(1)
4.	*Courtoisia* (a:2,b:1,c:0,d:0,e:2)		5(1),20(1)
5.	*Cyperus* X=8,9 (a:104,b:67,c:1,d:60,e:270)	2X,4X,6X,8X,10x, 12X,14X,16X,18X, 20X,22X,26X,28X	5(1),8(7),9(3),10(1),11(2),12(3),13(7),15(2), 16(7),17(3),18(12),19(5),20(7),21(6),22(1), 23(2),24(6),25(10),26(1),27(2),28(2),29(4), 30(1),32(4),34(2),35(1),36(10),37(2),38(1), 39(2),40(8),41(1),42(7),43(6),44(5),45(2), 46(2),47(5),48(13),49(7),50(4),51(1),52(6), 53(1),54(18),55(1),56(11),57(2),58(4),59(1), 60(5),62(6),63(3),64(11),66(1),67(1),68(1), 69(1),76(2),80(1),86(1),98(1),104(5),110(1), 112(1)
6.	*Eleocharis* X=5,8,9 (a:15,b:12,c:2,d:6,e:38)	2X,4X,6X,16X, 22X,24X	5(6),7(1),8(3),10(14),15(1),23(1),27(5), 38(2),40(1),>100(2),108(2)

1	2	3	4
7.	*Eriophorum* X = 5,9,19,29 (a:3,b:2,c:0,d:2,e:2)	10x	25(1),26(1)
8.	*Fimbristylis* X = 5,6,8,9,11 (a:55,b:36,c:20,d:7,e:145)	2X,4X,6X,8X	3(4),5(75),10(32),11(6),12(1),15(4),16(5), 20(6),22(2),24(1)
9.	*Fuirena* X = 5,29 (a:6,b:5,c:0,d:5,e:7)	8X,10X	18(2),119(4),26(1)
10.	*Kobresia* X = 26,29,36 (a:1,b:1,c:0,d:1,e:1)	-	40(1)
11.	*Lipocarpha* X = 23 (a:2,b:2,c:0,d:2,e:10)	-	13(6),16(1),19(3)
12.	*Queenslandiella* (a:1,b:1,c:0,d:1,e:1)	-	56(1)
13.	*Rhynchospora* X = 8,13 (a:9,b:2,c:0,d:2,e:2)	2X	9(1),10(1)
14.	*Rikliella* (a:2,b:1,c:0,d:1,e:1)	-	20(1)
15.	*Schoenus* X = 9,11,19,27,38 (a:1,b:1,c:0,d:1,e:1)	4X	22(1)
16.	*Scirpus* X = 7,8,13,20,21 (a:24,b:20,c:1,d:19,e:55)	2X,6X,8X,10X, 12X,14X,16X	5(1),10(1),11(3),14(8),15(3),16(4),20(2),21(5), 22(3),37(2),38(1),39(1),40(3),42(1),44(3), 47(1),48(2),49(2),50(1),51(1),52(2),55(2), 56(1),57(1),58(1)

* a - number of taxonomically known species, b-number of cytologically known species, c-number of diploid species, d-number of polyploid species, e-total number of cytotypes/chromosomal races.

C. squarrosus 4x, 10x, 12x based on x=9; *C. tenuispica* 2x, 4x based on x=8; *C. triceps* 4x, 12x, 16x; based on x=8; *Eleocharis atropurpurea* 2x, 4x based on x=5; *E. chaetaria* 2x, 4x based on x=5; *E. dulcis* 8x, 22x, 24x based on x=9; *E. geniculata* 2x, 4x, 6x, 16x based on x=5; *E. plantaginoides* 22x, 24x based on x=9; *Fimbristylis cymosa* 2x, 6x, 10x based on x=5; *F. dichotoma* 2x, 4x, 6x based on x=5; *F. falcata* 2x, 4x based on x=11; *F. ferruginea* 2x, 4x, 6x based on x=5; *F. ovata* 2x, 4x based on x=5; *F. spathacea* 2x, 4x based on x=5; *F. tetragona* 2x, 4x based on x=5; *Scirpus lateriflorus* 2x, 4x, 16x based on x=5; *S. maritimus* 12x, 14x based on x=8; *S. roylei* 4x, 12x based on x=7, *S. tuberosus* 12x, 14x based on x=8 and *Carex baccans* 10x, 16x based on x=5

From the foregoing account it is evident that:

i) Indian sedges show high incidence of polyploidy (72.4%) as compared to the world figures given by Grant (1963) for the flowering plants as Dicots: 43.0% and Monocots 58%.

ii) Incidence of polyploidy is still higher in *Carex, Fuirena, Scirpus* and *Cyperus* species.

iii) 25 (14.7%) species lack any evidence of polyploidy.

iv) 22 (12.9%) species have both diploid as well as polyploid taxa.

v) There is no general lack of higher grades of polyploidy amongst Indian sedges except for the genus *Fimbristylis* where only 2x to 8x species are recorded.

vi) As many as 50 (29.4%) species show incidence of infraspecific polyploidy.

c) **Euploidy**

The role of euploidy in the evolution of the family Cyperaceae is not so much significant since the polyploids are predominantly aneuploids. However, it is interesting to note that the members of genus *Fimbristylis* generally exhibit euploidy. A perusal of world-wide data shows that 42.5% of the species of this genus show euploid series and dominance is at tetraploid level.

In the Indian sedges, euploid series have been noticed for:

a)	*Cyperus brevifolius*	n=16 (4x), 48 (12x), 56(14x)
b)	*C. compressus*	n=48 (12x), 56 (14x), 64 (16x)
c)	*C. difformis*	n=9 (2x), 18 (4x)
d)	*C. iria*	n=36 (8x), 54 (12x)
e)	*C. kyllingia*	n=48 (12x), 56 (14x)
f)	*C. rotundus*	n=48 (12x), 56 (14x), 64 (16x), 80 (20x)
g)	*Eleocharis atropurpurea*	n=5 (2x), 10 (4x)
h)	*E. chaetaria*	n=5 (2x), 10 (4x)
i)	*E. geniculata*	n=5 (2x), 10 (4x), 15 (6x), 40 (16x)
j)	*Fimbristylis dichotoma*	n=5 (2x), 10 (4x)
k)	*F. falcata*	n=11 (2x), 22 (4x)
l)	*F. ferruginea*	n=5 (2x), 10(4x)
m)	*F. ovata*	n=5(2x), 10(4x)
n)	*F. polytrichoides*	n=5(2x), 10(4x)
o)	*F. podocarpa*	n=10(4x), 15(6x)
p)	*F. quinquangularis*	n=5(2x), 10(4x)
q)	*F. squarrosa*	n=5(2x), 10(4x)
r)	*F. spathacea*	n=5(2x), 10(4x)
s)	*F. tetragona*	n=5(2x), 10(4x)
t)	*Scirpus tuberosus*	n=48(12x), 56(14x)

Among polyploid species only 13.8% are euploid.

d) **Aneuploidy**

Aneuploid numbers in Cyperaceae are of frequent occurrence in various taxa of the family. Stebbins (1971) stated that Cyperaceae is characterised by extensive aneuploid series and these are best known in the genera *Carex* and *Eleocharis*. Earlier, he (Stebbins, 1950) had suggested that high aneuploid series of chromosome numbers found in Cyperaceae and Juncaceae are probably modified forms of polyploidy.

Highest frequency of aneuploidy has been noticed in *Carex* where almost all the numbers from 2n=12 to 2n=112 are met with. It was Tanaka (1949), who first of all, pointed out the presence of races with aneuploid numbers in this genus. In *Eleocharis palustris* and *E. uniglumis*, Strandhede (1958) observed a large number of aneuploid and polyploid cytotypes. For *Scirpus* chromosome number between 2n=10 to 2n=164 have been counted with various aneuploid numbers. Similar is the case for *Eriophorum* with 2n=26 to 2n=76.

Previous records for *Cyperus* indicate widely varying chromosome numbers eg. 2n=16, 18, 20, 26, 224, etc. in different species showing clearly the excistence of extensive aneuploidy. Aneuploidy is otherwise absent in *Fimbristylis*, a most primitive member of the family but Ramakrishnan & Ramachandran (1976) have postulated that n=3 in *F. woodrowii* is a result of aneuploid reduction from n=5, a very common number for the genus.

Amongst Indian sedges aneuploid numbers have been recorded for 99 species as follows:

Bulbostylis capillaris	n=36
B. densa	n=32
Carex baccans	n=23,42
C. filicina	n=21,24
C. flacca	n=38
C. folioisa	n=29
C. glaucina	n=23
C. insignis	n=24
C. longipes	n=21
C. longicrusis	n=26
C. melonantha	n=24
C. fedia	n=47,52
C. muricata	n=28
C. nubigena	n=52
C. remota	n=31
Cyperus alopecuroides	n=23,44,49,50,52,54
C. amabilis	n=19
C. alulatus	n=57
C. brevifolius	n=60,62,63,86
C. bulbosus	n=46,50,57
C. cephalotes	n=21
C. castaneus	n=20,49
C. compactus	n=12,21,43,47
C. compressus	n=49
C. corymbosus	n=54
C. cyperoides	n=41,44,67,110
C. cyperinus	n=68,76
C. diaphanus	n=25
C. difformis	n=13,16,17
C. diffusus	n=19

C. digitatus	n=42,52
C. distans	n=27,29
C. elegans	n=25
C. eleusinoides	n=52
C. esculentus	n=104
C. flabelliformis	n=15
C. flavescens	n=35
C. flavidus	n=10,36
C. fuscus	n=36
C. globosus	n=36,42,43,44
C. haspans	n=13,18
C. imbricatus	n=55
C. inundatus	n=5,29,30
C. iria	n=52,58,64
C. kyllingia	n=12,47,51,60,62
C. laevigatus	n=42,43,44
C. latovaginata	n=12
C. michelianus	n=38,39
C. niveus	n=38,39
C. nitens	n=47
C. nutans	n=34,36,37
C. odoratus	n=25
C. paniceus	n=42,98
C. pilosus	n=34
C. platystylis	n=20
C. plumbonucea	n=11
C. polystachyus	n=49,53,64
C. procerus	n=13,32,52
C. pseudokyllingioides	n=20
C. rotundus	n=13,40,42,50,52, 55,58,66,69,74
C. rubicundus	n=21,25,26
C. sanguinolentus	n=25
C. squarrosus	n=21,47,49
C. substamineus	n=21

C. tenuiculmis	n=22
C. thomsoni	n=28
C. triceps	n=11,46,49,62
C. welwitschii	n=62
Eleocharis dulcis	n=38, >100
E. plantaginoides	n= >100
E. quinqueflora	n=23+0-3B
Eriophorum comosum	n=26
Fuirena cilliaris	n=19
F. trilobites	n=19
F. umbellata	n=26
F. uncinata	n=18
F. wallichiana	n=19
Fimbristylis cymosa	n=16,24
F. bisumbellata	n=8
F. dichotoma	n=16
F. ferruginea	n=16
F.umbellaris	n=3
F. narayanii	n=3
F. woodrowii	n=3
F. squarrosa	n=12
Rhynchospora corymbosa	n=9
R. wightiana	n=10
Scirpus articulatus	n=10,15
S. erectus	n=37
S. grossus	n=44
S. isolepis	n=15
S. juncoides	n=38
S. kyllingioides	n=50,51,52
S. lateriflorus	n=5,11,15
S. maritimus	n=55
S. planifolius	n=22
S. squarrosus	n=20
S. roylei	n=11,44
S. tuberosus	n=47,49,57,58

Jones (1978) suggested centric fissions and fusions as the main factor responsible for aneuploidy in Cyperaceae with diffuse nature of the centromere. Earlier, for *Eleocharis*, Hoakansson (1954) suggested that fragmentation and perhaps also fusion have contributed to more familiar changes; autopolyploidy and true aneuploidy. He further stated that formation of new chromosome by fragmentation is possible in chromosomes with diffuse or several centromeres.

In *Carex*, Heilborn (1939) described non-conjunction, bastardization with subsequent segregation of aneuploid numbers, structural changes of chromosomes and autopolyploidy leading to instability of chromosome numbers. Earlier, allopolyploidy was also supposed to be responsible for the aneuploidy but it was ruled out by Heilborn (1939).

Different persons have given quite variable explanations for the possible cause of aneuploidy. Navashin (1930) attributed the phenomenon to abnormal spindle activity, whereas Narayanan (1951) correlated it to the duplication of one or two chromosomes of a complement. According to Richardson (1933), Mac Mohan (1936) and Sachdeva (1972), aneuploidy is the result of fragmentation of chromosomes but unequal reciprocal translocations are considered as the cause of aneuploidy by Togby (1943), Babcock (1947) and Stebbins (1950, 1971, 1974). At the same time erroneous counts and wrong identifications of the materials under the same taxonomic unit cannot be overlooked. However, this should not minimise the role of aneuploidy in the evolution.

e) **Polyploidy and Habit** Correlation

The origin and success of polyploids quite often depends upon the habit, habitat as well as the breeding system (deWet 1980). Muntzing (1936), Stebbins (1938, 1950, 1971), Gustaffson (1947, 1948) and Faverger (1967) are of the opinion that polyploidy is more common among perennial herbs than annual and woody species. Baquar (1976) while compiling the flora of Pakistan concluded that the incidence of polyploidy is high in perennials as compared to annuals. For Indian plants, mention may be made here of studies of Bir & Sidhu (1983) on weed flora of Punjab. They found that in cultivated fields the annual polyploids are more stabilized and grow better than the perennial polyploids. But Sahni (1987) for Punjab grasses found that percentage of polyploidy is nearly the same amongst annuals and perennials. Cheema (1991) while studying sedges of NW India found that the percentage of polyploids is more in perennials and this is in accordance with Baquar's (1976) surmise for Pakistan flora.

But, Wright (1967) rejected the concept of polyploidy-habit correlation. The origin of perennial habit as well as accessory method of vegetative reproduction are considered to be a direct consequence of polyploidy (Muntzing, 1936; Gustaffson 1947, 1948).

ACKNOWLEDGEMENTS

We wish to express our gratitude to UGC (Paramjit Cheema as Research Associate) and CSIR (S.S. Bir as Emeritus Scientist), New Delhi for financial assistance.

REFERENCES

Babcock, E.B. (1947): *Uni. Calif. Publ. Bot.* **21 & 22**: 1030.

Baquar, S.R. (1976): *Taxon* **25**· 621-627.

Bedi, Y.S., Gill B.S. and S.S. Bir (1985): *J. Cytol. & Genet.* **20**: 162-203.

Bir, S.S. (ed.) (1983-1993): *SOCGI Chromosome Number Reports I-XI. Jour. of Cytol. & Genet.*

Bir, S.S. and P. Cheema (1994): Communicated to *Cyperaceae Newsletter* (Kew).

Bir, S.S. and M. Sidhu (1983): In: R.R. Sinha and U. Sinha (eds.) *Curr. Approaches in cytogenetics*. Spectrum Publ. House, New Delhi, pp. 135-144.

Bir, S.S., M. Sidhu and Kamra (1986): *Cytologia* **51**(1):95-106.

Bir, S.S. Sindu; Karma, M. and P. Cheema (1988a): *J. Cyto. and Genet.* **23**:14-37.

Bir, S.S., Cheema, P. and M. Sidhu (1988b): *J. Cytol. and Genet.* **23**:220-221.

Bir, S.S., Cheema, P. and M. Sidhu (1990): *The Nucleus.* **33** (1,2):22-24.

Bir, S.S., Cheema, P. and M. Sidhu (1991): *Cytologia* **56** (1,2):645-651.

Bir, S.S., Cheema, P. and M. Sidhu (1992a): *Proc. Ind. Nat. Sci. Acad. Biol. Sci.* **B58**:63-70.

Bir, S.S., Cheema, P. and M. Sidhu (1992b): *J. Pl. Sc.* **8**:4-8.

Bir, S.S. Chatha, G. S. and M. Sidhu (1992c): *Willdenowia* **22**:133-142.

Bir, S.S., Cheema, P. M. Sidhu and S. Kumari (1993a): Ind. *Nat. Sci. Acad. Biol. Sci.* **59**:163-173.

Bir, S.S., Cheema, P., Kumari, S. and M. Sidhu (1993b): *Curr. Sci.* **66**:322-324.

Bir, S.S., Cheema, P. and C.P. Singh (1993c): Communicated to *Bangaladesh J. Bot.*

Briggs, D. and S.M. Walter (1994): *Plant varitions and evolution,* Cambridge University Press Cambridge.

Cheema, P. (1994): *Cytomorphological studies on members of family Cyperaceae from Punjab plain and adjoining regions.* Ph.D. thesis Punjabi University , Patiala.

Cheema, P. and S.S. Bir (1994): In : *Proceedings of 4th All India conference on Cytology and Genetics* Banglore on 5-7 November (Accepted).

Cheema, P. and S.S. Bir (1995): In : *Perspectives in Cytology and Genetics* **8** (Accepted).

Cheema, P. Kumari, S. Sidhu, M. and S.S. Bir (1992a): *Cyto and Genet.* **27**:163-173.

Cheema, P. Bir, S.S. and M. Sidhu (1992b): *Nucleus* **35**:83-86.

Cheema, P. and S.S. Bir (1993a): *Cyperaceae News Lett.* **12**:5-7.

Cheema, P. Bir, S.S. and M. Sidhu (1993b): *Cytologia* **58**:345-349.

Cave, M.S. (ed.) 1958-60: *Index to Plant Chromosome Numbers previous to 1959 and for 1956-1959,* and Suppl. Vol. published by California Bot. Soc., California.

Cave, M.S. and Chisski-Hommersand (eds.) 1961-65. *Index to Plant Chromosome Numbers for 1960-1964*. University North Carolina Press, North Carolina.

Chatha, G.S. and Bir, S.S. (1986): *J. Cytol. and Genet.* **21**:97-114.

Darlington, C. D. (1965): *Cytology*, J. and A. Churchill Ltd., London.

Darligton, C.D. (1973): *Chromosome, Botany and the origin of Cultivated Plants.* George Allen and Unwin Ltd.

Darligton, C.D. and A.P. Wylie (1955): *Chromosome Atlas of Flowering Plants.* George Allen and Unwin Ltd., London.

Davies, P.H. and V.H. Heywood (1963): *Principles of Angiosperm Taxonomy,* London.

DeWet, J.M.J. (1980): In : W.H. Lewis (ed.) *Polyploidy, Biological Relevance.* Plenum Publishing Corp., New York. pp 3-15.

Favarger, C. (1967): *Biol. Rev.* **42**:163-206.

Favdrov, AN. A. (ed.) (1969): *Chromosome numbers of flowering plants.* Academy of Sciences of the USSR.

Gustafsson, A. (1947): *Lunds University, Aarsskr.* **44**:188-370.

Gustafsson, P. (1980): *Hereditas* **34**:1-22.

Goldbatt, P. (1980): In: W.H. Lewis (ed.) *Polyploidy, Biological Relevance*. Plenum Publishing Corp. New York. pp. 219-239.

Goldblatt, P. (ed.) (1981-1985): *Index to Plant chromosome numbers for 1975-1983.* Missouri Botanical Garden.

Grant, V. (1963): *The origin of Adaptations.* Columbia University Press, New York.

Hakansson, A. (1954):*Hereditas* **40**:325-345.

Heilborn, O. (1939): *Hereditas* **25**:224-241.

Jones, K. (1978): In : H.W. Woodhouse (ed.) *Advances in Bot. Res.* **6**:120-191.

Khosla, P.K. (1978): *Nucleus* **21**:211-218.

Kumar, V. and B. Subramaniam (1987): *Chromosome Atlas of Flowering Plants of Indian Sub-continent.* BSI.

Love, A. (1972): *Taxon* **21**:679-684.

Love, A. and D. Love (1974): *Cytotaxonomical Atlas of the Slovenian Flora. J. Cramer,* 3301 Lehre.

Love, A. and D. Love (1975): *Cytotaxonomical Atlas of Arctic Flora.* J. Cramer, Fl-9490 Vaduz.

Love, A. and D. Love/ (1980): *Taxon* **30**:707-709.

Love, A. and D. Love (1991): *Taxon* **30**:845-851.

Love, A. and Love, D. and A., Raymond (1957): *Jour. Bot.* **35**:715-761.

Machmohan, B. (1936): *Cellule* **45**:209-262.

Mehra, P.N. (1982): *Cytology of east Indian grasses*. New Delhi.

Mehra, P.N. and S.K.Sachdeva (1975a): *Taxon*. **24**:501-516.

Mehra, P.N. and S.K. Sachdeva (1975b): *Cytologia* **40(3/4)**:497-515.

Mehra, P.N. and S.K. Sachdeva (1976): *Cytologia* **40(3/4)**:585-590.

Moore, R.J. (1974): *Regum Vegetabile,* **91**, Utrecht, Netherland.

Moore, R.J. (ed.) (1977): *Regnum Vegetabile,* **96**, Utrecht, Netherland.

Muntzing, A. (1936): *Hereditas* **21**:263-328.

Narayanan, C. R. (1951): *J. Madras Univ.* **21B**:220-231.

Navashin, M. (1930): *Univ. Calif. Pub. Agri. Sci.* **6**:95-106.

Nijalingappa, B.H.M. (1973): *Taxon* **22**:459-464.

Nijalingappa, B.H.M. (1974): *Proc. Indian Acad. Sci. LXXX.* No. **3**:134-138.

Nijalingappa, B.H.M. (1975): *Cytologia* **40(1)**:177-183.

Nijalingappa, B.H.M. (1977): *Proc. Indian. Acad. Sci.* **85B No. 1** pp. 21-24.

Nijalingappa, B.H.M. and P. Leela Bai (1990): *Cytologia* **55(3)**:373-379.

Nijalingappa, B.H.M., Nagraj, N. and D., Tejavathi (1978): *Taxon* **27**:529-535.

Ornduff, R. (ed.) (1967-68): *Index to Plant chromosome Numbers for 1965-66* Int. Bur. Plant Taxonom. Numenclature, Netherlands.

Patnaik, S.N. and S.P. Rath (1983): *Curr. Appr. in Cytogenetics,* 123-133.

Ramakrishnan, S. and K. Ramachandran (1976): *Nucleus* **19(1)**:20-25.

Rath, S.P. and S.N. Patnaik (1974): *Cytologia* **39 (2)**:341-352.

Rath, S.P. and S.N. Patnaik (1977): *Bot. Mag.* **90**:79-81.

Rath, S.P. and S.N. Patnaik (1978): *Proc. 65th Indian Sci. Congr., Part 3* :107-108.

Rath, S. P., Mohapatra, J. and S. N. Patnaik (1973): *Proc. Indian. Sci. Congr. Assoc.* **60**:314-316.

Richaradson, M.M. (1933): *Uni. of Calif. Publ. Bot.,* **17**:51-60.

Rogera, D.J. and S.G. Apan, S.G. (1969): *Taxon* **18**:609-624.

Sachdeva, S.K. (1972): *Cytomorphological studies on West Himalayan monocots with particular reference to species of economic importance.* Ph.D. thesis, Panjab University , Chandigarh.

Sahni, M. (1987): *Cytotaxonomy of Punjab grasses.* Ph.D. thesis, Panjabi University, Patiala.

Sanyal, B. and A. Sharma (1972): *Cytologia* **37**:13-32.

Santhapau, H. and A.W. Henery (1973): *A dictionary of the flowering plants of India.* Pub. and Infor. Directorate, CSIR, New Delhi.

Sarkar, A. K., Chakraborty, M., Saha, N.C. and S.K. Das (1976): *Taxon* **25**:631-649.

Sharma, A.K. and A.K. Bal (1956): *Proc. 41st Indian Sci. Congr. Part 3*:147.

Stebbins, G.L. (1938): *Amer. J. Bot.* **25**:189-198.

Stebbins, G.L. (1950): *Variations and Evolutions in plants.* Columbia University Press, New York.

Stebbins, G.L. (1971): *Chromosomal Evolution in Higher Plants.* Edward Arnold Publ., London.

Stebbins, G.L. (1974): *Flowering Plants Evolution Above the species Level.* Edward Arnold publ., London.

Strandhede, S.O. (1950): *Skandinavian and Finland Bot. Not.* **111**:228-236.

Subramanian, D. (1988): *Cytologia* **53(1)**:67-72.

Swanson, C.P. (1957): *Cytology and Cytogenetics.* Prentice Hall, Inc. Engle wood Cliffe, N. J.

Tanaka, N. (1937): *Cytologia, Fujii, Jub. Vol.* 814-821.

Tanaka, N. (1940): *Cytologia* **10**:348-362.

Tanaka, N. (1948): *Biol. Contrib. Japan* **4**:1-327.

Tanaka, N. (1949): *Cytologia* **15**:15-29.

Tejavathi, D.H. (1988): *Curr. Sci.* **57**:724-728.

Tobgy, H.A. (1943): *Jour. Genetics.* **45**:67-111.

Walker, H.I. (1938): *J. Arn. Arb.* **19**:158-162.

Wahl, H.A. (1940): *Am. Jour. Bot.* **27**:458-470.

Willis, J.C. (1966): *A Dictionary of flowering plants and ferns.* 7th ed. Revised by Airyshaw. Cambridge University Press, London.

Wright, J.W. (1967): *F.A.O. forestry and Forest product studies.* **16**:399.

Wulff, H.D. (1939): *Jahrb. Wiss. botanik,* 87.

Taxonomy of Bacterial Plant Pathogens

H.C. Dube and N.H. Zadeh*

Department of Life Sciences,

Bhavnagar University,

Bhavnagar 364 002.

A century has elapsed since Burrill (1882) unmistakably and unequivocally demonstrated the bacterial etiology of plant diseases, but even today phytobacteriology compares poorly with other areas of bacteriology like medical or animal bacteriology. The causes are not far to see. According to Starr (1983) it is because of disciplinal insularity. The phytopathogenic bacteria were studied by plant pathologists in facilities separate from those devoted to other areas and there was no intellectual interaction. Their papers were published in different periodicals; they attended separate scientific meetings and the paths of phytobacteriologists, animal pathologists and bacteriologists seldom crossed. Although the barriers have breached in recent years, they have by no means disappeared.

It is not surprising, under these circumstances of disciplinal insularity, that the early bacterial taxonomists practiced a nomenclatural insularity by erecting separate taxa for the phytopathogenic bacteria, segregating them from all other bacteria solely on the basis of plant pathogenicity. For example the first edition of "Bergey's Manual of Determinative Bacteriology" placed all the phytopathogenic bacteria separately in a tribe Erwinieae, defined simply as "Plant pathogens" (Bergey *et al.*, 1923).

---

* Present address: Plant Pests and Diseases Research Institute, P.O. Box 1454, Tehran 19395, IRAN.

The "Erwinieae notion" remained with us for half a century. Happily, the situation is fast changing and with the publications of some invaluable laboratory manuals and books and treatises phytobacteriology is poised for a big leap.

In recent years, studies with plant pathogenic bacteria have been extended to bacteria that are not causing diseases but are in juxtaposition with them on the surface of plants or in their vicinity. What is their role in the life of phytopathogenic bacteria and in the infection process ? We know much about bacteria that share the rhizosphere with the plant pathogens during their saprophytic phase, but we have little information about those that live as epiphytes on the surface of the plants. According to Sands *et al.* (1982), phytogenic bacteria are those epiphytes that have escaped from microbial competition on the plant surface. It will be rewarding to discover the relationship (not interaction, as this term is preferably used between unrelated organisms) between epiphytic, saprophytic, and parasitic bacteria on the 'aerosphere' (aerial plant surface). Last and Deighton (1965) demonstrated the existence of saprobic, usually pigmented bacteria, on the leaves (phylloplane) in greater amounts than in the soil that constituted a characteristic ecological group. Since then, a wide and interesting variety of microorganisms have been reported on the plant surfaces, majority of them are gram-ve bacteria like *Erwinia, Flavobacterium, Pseudomonas* and *Xanthomonas.* The gram +ves are represented by *Bacillus, Corynebacterium,* and *Lactobacillus.* These epiphytic bacteria may effect the pathogenicity of plant pathogens. Some of them may even enter the plant and elicit responses from the plant that may have a bearing on the host-pathogen relationship. The two most common epiphytic bacteria are *Erwinia herbicola* and *Pseudomonas flourescens.* Similarly, the epiphytic spiroplasmas (plant mycoplasmas) have no pathogenic potentials and colonise only surface of flowers (Davis, 1979, 1981). Among the intriguing questions proved by the discovery of spiroplasmas in this specialized ecological niche, are those related to their origin, relationships with other spiroplasmas, the nature of their vectors and hosts, and with other microorganisms at the flower surface. The flower spiroplasma have become an enigma.

It appears that the resident epiphytic bacteria occupying the phylloplane, contribute towards protection against diseases. When the population is large, some nonpathogenic bacteria (along with the pathogens) enter the plant. Some enter actively beforehand and provide "protection." The entry induces host responses similar to phytlolexin production. Nature may be operating biological control through epiphytes. The line between pathogenic and saprobic bacteria seems thinning down. Genus *Erwinia* includes many saprobes, and *E. coli* has been related to *Erwinia amylovora, E. chrysanthemi* and *E. herbicola.* Verma *et al.* (1980) have reviewed several important aspects in this area some strains of *E. herbicola* when incubated at a temperature slightly above optimum for growth (34°), changed irreversibly to non-pigmented, auxotrophic (for thiamine) variant having closer serological relationship with pathogenic *E. amylovora* group. Such transformations occurring under field conditions will surely necessitate more serious examination of the epiphytic organisms.

Naming the bacteria on the host basis has to go. Pathogenicity has been rightly been challenged as the reason to segregate them, as pathogenicity is not a 'head' or 'tail' affair. It is conditioned by several factors disease-causing isolates as 'new' on few characters, mostly on the host, without making comparisons with other bacteria, is no more tenable. Similarly, the usual emphasis on 'proving' uniqueness of the 'new' species has led to chaos of probable synonymy. Thus, the "new host-new species" cliche is rejected. Many plant-pathogenic bacteria, including most pseudomonads and xanthomonads, were relegated to pathovars (pv) from species level by Dye *et al.* (1980). The 'patho-species' concept so firmly entrenched, has been accommodated under this intraspecific designation pv.

Plant pathologists lack a sound reference system to which the unknown isolates could be compared. They have been denied a taxonomy which is ideally based on a classification which groups together closely related strains, a logical nomenclature, and a realistic identification system as investiged by Williams (1985). At a meeting of Judicial commission of the International Commission of Systemic Bacteriology (ICSB) in 1973, it was decided to produce a review of the recently valid names of bacteria, with the object of retaining only those taxa that were adequately described and for which there was a type, a neotype or reference strain available. This was an extremely useful and timely project that resolved many nomenclatural anomalies and provided a baseline for future designations of new taxa. Skerman *et al.* (1980) have listed the plant pathogenic bacteria approved by the judicial commission of the ICSB (Table I).

The new list is on lines of the VIII edition of the Bergey's Manual (1974). In absence of complete descriptions of the species, many of them were (unfortunately) reduced to the level of pvs and several were totally eliminated e.g., *Streptomyces scabies.* Schaad (1982) feels that it was unfortunate, because many of these pathovars are legitimate species. The current list of approved names contains only three phytopathogenic fluorescent pseudomonads *viz. P. syringae, P. cichorii* and *P. viridiflava.* All other fluorescent pseudomonads listed in the VIII edition of Bergey's Manual are currently considered invalid due to lack of data and, therefore termed pathovars of *P. syringae.* On the contrary. DNA homology data are available, which indicate differences among phytopathogenic pseudomonads and does not support the "pathovar" taxomomy. Schaad (ibid) says "Accepting the indiscriminate use of pathovar for many valid species is a mistake. This is not only a rejection of pathogenicity as an important character but also a rejection of other important data. Certainly many plant pathogenic bacteria have been poorly characterized and do not deserve species ranking, but many currently designated as pathovars have been well characterized. Furthermore, many xanthomonads and fluorescent pseudomonads are easily differentiated not by traditional physiological tests, but by such methods as DNA homology, serology-PAGE, and nutrition. The use of "pathovar" certainly has its place in the nomenclature of plant pathogenic bacteria. On the other hand, its use should be restricted to those organisms that can not be differentiated by other traits.

The Genera

The plant pathogenic genera are placed in the VIII edition (1974) of Bergey's Manual in various parts as given below:

Part 7:	*Pseudomonas, Xanthomonas, Agrobacterium*
Part 8:	*Erwinia*
Part 15:	*Bacillus, Clostridium*
Part 17:	*Corynebacterium, Streptomyces*
Part 18:	*Rickettsia*- like organisms
Part 19:	*Spiroplasma* (plant mycoplasma)

In the latest Bergey's Manual (1984), renamed as "Bergey's Manual of Systematic Bacteriology" (4 volumes), the plant pathogenic bacteria have been classified in various sections as follows:

Section 4:	*Pseudomonas, Xanthomonas, Agrobacterium*
Section 5:	*Erwinia*
Section 6:	*Rickettsia*- like organisms
Section 10:	*Mycoplasmas*
Section 13:	*Bacillus*

Section 15: *Corynebacterium*

Section 17: *Streptomyces, Nocardia*

Genus Pseudomonas

It includes the most versatile bacteria, colonizing all sorts of habitats and some are plant pathogenic. There are suggestions to make at least five genera from among the present species. The various species fall into two broad categories of fluorescent and non fluorescent pseudomonads. The former do not accumulate PBHB (poly ß-hydroxybutyrate) while the latter do. But both, fluorescence and PBHB accumulations are not reliable. Starr (1983) said like this: "The vagaries of *Pseudomonas* taxonomy have led some to conclude that identification is an ill-defined,some- what chaotic mystic art".

Skerman *et al.* (1980) have recognized only three species in the fluorescent pseudomonas group viz.: *P. cichorii* and *P. viridiflava*. All other species listed in the VIII edition of Bergey's Manual stand invalid and are recognized as pathovars of *P. syringae* Starr. (1981) has recognized 19 nomen species in *P. syringae;* some important ones being *P. savistoni* (incites tumors on Oleander and Olive), *P. phaseolicola, P. coronofaciens var. autropurex, P. glycinea, P. morsprunorum* (all causing chlorosis).

Majority of the *Pseudomonas* species are non-fluorescent: *P. solanacearum* being the most important causing several wilt diseases in the tropics. *P.gladioli (=P. marginata)* causing rot of Gladiolus now includes *P. allicola,* the onion-rot pathogen. *P. caryophilli* also causes rots.

Xanthomonas

This is a genus that comprises exclusively of plant pathogens. It forms typical dome-shaped, yellow, highly mucoid colonies on sugar-containing media. YDC (Yeast extract Dextrose Calcium carbonate) medium is the best selective medium in which almost all xanthomonads grow at 25-30°C. The typical yellow pigment and the slime constitute the two diagnostic traits of *Xanthomonas.* Both are unique. The yellow pigment is not a carotenoid like the ones produced by *Erwinia, Pseudomonas,* or *Corynebacterium,* but a mixture of brominated aryl-polyene esters called Xanthomonadins. Crude extracts of the pigment absorb light maximally at 445 nm and provide most reliable identifications of xanthomonads.

The slime, responsible for the mucoid nature of the colonies, is a heteropolysaccharide, called xanthan gum. It is made up of a backbone of ß-1,4 linked glucose units with short side chains comprising of mannose and glucuronic acid units. Being economically valuable, it is manufactured from *X. campestris.*

Only five species that can be differentiated by physiological tests are valid. These are *X. albilineans, X. ampelina, X. axonopoids, X. campestris* and *X. fragariae.* Species like *X. oryzae, X. vesicatoria, X. malvacearum, X. transluscens* etc. are now pathovars of *X. campestris.*

Agrobacterium

The agrobacterium are common soil inhabitants and constitute a common component of soil microflora especially in the rhizosphere. These have long periods, 2-3 years, of survival in the soil. The non-phytopathogenic forms predominate in soils and even in the vicinity of the diseased plants. Only about 1% of the isolates from soil are phytopathogenic which produce copious slime on carbohydrate-containing media. Colonies are non-pigmented and smooth, though some produce rough colonies.

Three pathogenic species are: *A. rhizogenes, A. rubi* and *A. tumifaciens.* All of them induce tumorous outgrowths.*A. rhizogenes* causes proliferations or hairy outgrowths on the secondary roots. *A. rubi* causes galls on aerial parts of canes while *A. tumirfaciens* causes galls, " crown gall," on dicotyledonous plants (only). *A. radiobacter*, a nonpathogenic species is used to control *A. tumirfaciens* (biological control).

Erwinia

It includes plant pathogens as well as epiphytic organisms of plant surfaces while some are opportunistic pathogens of man and animals. *Erwinia* is placed in family Enterobacteriaceae along with *Escherichia, Shigella, Salmonella, Proteus* and *Serratia* with whom its close relationship is recongnised since the beginning of the century.

The species are placed in three 'natural' groups viz: *carctvora, amylovora* and *herbicola* group.

The *carotavora group* of organisms cause soft-rots and show strong pectolytic activity and once upon a time a separate genus, *Pectobacterium,* was named for these organisms. *E. carotovora* var. *carotovora* (Ecc) and *E. carotovora* var. *atroseptica* (Eca) contain several strains many of which are identifiable serologically. The soft-rot of carrot is caused by Ecc while the 'black leg' of potato is caused by Eca.

The *amylovora group* are 'true erwinias'; *Erwinia sensu stricto.* It is typified by *E. amylovora,* the agent that causes the 'fireb-ght' of apple and with which phytobacteriology was started in the last quarter of the 19th century. The members of this group are nonpigmented and nonpectolytic and cause dry necrosis or wilt symptoms. *E. quercine, E. rubrifaciens* (canker of persian walnuts), *E. salicis* (wilt of willow) and *E. tracheophile,* causing wilt of cucumbers are the important species.

The *herbicola group* are also called 'yellow erwinias', but the pigment is a carotenoid and thus, distinguishable from the xanthomanadin of xanthomonads. The members grow on leaves and other parts as epiphytes. Those showing doubtful pathogenicity are placed under *E. milletiae, E. anans, E. lathyri, E. stewartii* etc. Erwinias isolated from man and animals are referred as "human clinical erwinias" or *E. hebicola* group from human sources, which are called *Enterobacter agglomerans.* The epiphytic erwinias have been reported to reduce the severity of the 'fire blight' of apples and pears and can be harnessed for biological control. Some members have been reported to fix nitrogen which explains their ability to colonize nitrogen deficient habitats like leaf surface.

Corynebacterium

Most of the species of *Corynebacterium* have been isolated from single hosts and named after the major symptoms that they induce. Dye (1974) has placed in five groups viz: *C. fascians, C. raythayi, C. sepedonicum, C. insidiosum* and *C. flacumfaciens.* Many of the species have been named pathovars of other species.

Streptomyces

It is filamentous bacterium forming non-motile spores in long chains of diagnostic value. The colonies are small, leathery, initially smooth-surfaced but later form abundant aerial hyphae which appear powdery or granular. The organisms produce a variety of melanoid pigments which alongwith the spore chain morphology, are helpful in identifying the species. *Streptomyces* spp. are most abundant in the soil and are the source of several antibiotics some of which are also used in the control of plant diseases.

Streptomyces causes scab and rots in a variety of root crops, the best known being the potato-scab which is caused by several species, though *S. scabies* is traditionally regarded as the causal organism. Ironically, this species is no more a valid species even though it is recognized by the International *Streptomyces* Project. Out of three species recognized earlier viz: *S. intemedius* (sugarbeet scab agent), *S. ipomeae* (sweet potato scab), and *S. setori* (one of the several, associated with potato scab), only *S. ipomeae* is now valid.

Nocardia

Nocardia also is filamentous but is different from *Streptomyces* in: (i) the fragmentation of its hyphae forming coccoid, elongate, non-motile spores, (ii) lack of aerial hyphae and sporangiophores, and (iii) the +ve acid fast reaction shown by *Mycobacterium. N. vaccinii* is the lone plant pathogenic species which causes galls and bud proliferations in blue-berry plants. But this disease has been reported only once in 1952 (See Vidave and Starr, 1981).

Rickettsias and Chlamydias (or Rickettsia- like organisms)

These organisms were unfamiliar to plant pathologists untill 1970 when Giannotti *et al.* described a 'bacterium-like' organisms in dodder. He, of course, did not think of rickettsias and it was Maramorosch, who in 1974, drew attention to this report and introduced rickettsias among plant pathogens. Much interest in these organisms was generated after discovery of rickettsia-like organisms (RLOs) from yellow group of diseases which had already yielded a new taxon, mycoplasmas, to plant pathogens. The widespread occurrence of RLOs in phytophagus homopterus insect prompted search for these organisms in deseased plants. Electron microphotographs showed microorganisms with a distinct, conspicuously rippled or rigid wall in contrast to the wallless, membrane bound mycoplasmas. These agents were sensitive to penicillin, resulting in remission of symptoms. (Mycoplasmas are not sensitive to penicillin).

Plant pathogenic rickettsias are confined to either xylem or phloem, and are segregated as xylem-limited, or phloem-limited walled RLOs. The habitats are not negotiable. Plant mycoplasmas inhabit only the phloem. The xylem-limited RLOs are spread by xylem-feeding insects like frog-hoppers and sharp-shooter leaf-hoppers that are yet unidentified.

The xylem-limited RLOs: The important diseases caused by these organisms are the Pierce's disease (PD) of grapewines, almonel leaf scorch (ALS) disease (these two agents are same and called PD-ALS bacteria), alfa-alfa dwarf phony peach disease (PPD), pulm leaf scald, plum leaf scorch, young tree decline of *Citrus* and perwinkle wilt. The xylem-limited RLOs induce typical wilt symptoms.

In electron microphotographs, the bacterial cells are found unaggregates immersed in electron dense matrix that are connected by fibrous strands. These strands are extra-cellular polysaccharides secreted by these bacteria which function in their survival and pathogenesis. The PD-ALS organisms have been taken in culture. It is transmitted by at least 26 species of leaf hoppers and 5 species of spittle bugs through a non-circulative mechanism. The PD-ALS agent, thus, has a wide host range infecting both monocots and dicots spread over 20 families. Leaf hoppers can acquire and transmit the RLOs in less than two hours and the infected adults remain infective for life. There is, however, no vertical transmission to the progeny and are themselves not adversely affected.

Phloem-limited RLOs: These include both rickettsias and chlamydias. The diseases caused by chlamydia-like organisms are: the clover club leaf, *Citrus* greening, and potato leaflet stunt.

The diseases caused by pholem-limited rickettsias are infectious and are necrosis of grapewine, rugose leaf-curl of clove, apple proliferation and 'little leaf' of *Sida cordifolia*.

The vector of phloem-limited rickettsias and chlamydias are not yet identified and in most cases the vectors are not known. Vertical transmission has been shown in leaf-hoppers transmitting the clove club leaf agent. Some of them have been taken in culture which will facilitate their identification. The symptoms induced by chlamydias include leaf stunting and clubing.The symptoms are mild and recovery is frequent.

The xylem or phloem limited organisms, other than the mycoplasmas, although referred as RLOs, are more appropriately termed as "eubacteria of unknown affinity". Without doubt future discoveries in these unusual habitats will be exciting.

Spiroplasma (Plant mycoplasmas)

It includes all the plant pathogenic mycoplasmas reported so far. The genus is characterized by: i) formation of helical motile filaments in the early growth phase, ii) susceptibility to tailed phages, and iii) Gram +ve reaction.

Athough most of the isolated spiroplasmas cause diseases in plants, their primary hosts are insects. Some spiroplasma, like the Sex Ratio Organisms (SRO), causing elimination of males from the progeny of *Drosophila*, and the Suckling Mouse Cataract Agent (SMCA), earlier believed to be caused by spirochetes, have been isolated only from insects and are not known to insect plants. Several 'bee spiroplasmas', lethal to honeybees, have been isolated from foraging bees which contract them from flowers harbouring epiphytic spiroplasmas.

The isolates of *Spiroplasma* show only minor variations in their physiological and biochemical properties, but they do show serological and distinctions on the basis of which they have been placed under seven groups. Nevertheless, at species level only one species has been recognized viz. *S. citri*. The plant mycoplasmas, i.e. spiroplasmas, have a life cycle that requires a period of multiplications in insect vectors *viz.* leaf hoppers, followed by transmission to the phloem of plants where they multiply and cause disease. Dependence on vectors for transmission, restricts the host range of spiroplasmas only to those plants which are visited by the insect vectors. Grafting between compatible plants also transmits the spiroplasmas.

Phytobacteriology is now a very fast-growing field of research and modern techniques of molecular biology like RFLP or genetic 'finger printing' in common parlance, RAPD, AP-PCR and AFCPs are now being routinely used in elucidating the identity and taxonomic 'distance' of bacterial plant pathogens. Plant pathologists or plant bacteriologists are no more using separate techniques. Disciplinal insularity is now a matter of history.

REFERENCES

Bergey, D.H., Harrison, F.C., Breed, R.S., Hammer, B.W. and F.M., Huntoon; F.M. (1923): *Bergey's Manual of Determinative Bacteriology, 1st ed. williams & Willkines* Baltimore.

Burrill, T.J. (1882): An account of the nature and effects, together with a systematic description of the species. *Illinoiis Industrial University, 11th Report :* 93-157.

Davis, R.E. (1979): Spiroplasmas: In *Leafhopper Vector and Plant Disease Agents.* (Maramorosch, K. and K.E. Harris, eds.) Academic Press. pp 451-484.

Davis, R.E. (1981): In *Mycoplasma Diseases of Trees and Shrubs* (Maramorosch, K. and S.P Raychaudhuri eds.). Academic Press, pp 259-279.

Dye, D.W. and Lelliot, R.A. (1974): In *Bergey's Manual of Determinative Bacteriology*.

Dye, D.W., Bradburry, J.F., Goto, M., Hayward, A.C., Lelliot, R.A. and M.N. Schroth (1980): *Rev. Plant Pathol.* 59:153-168.

Last, F.T. and F.C. Deighton (1965): *Trans. Brit. Mycol. Soc.* **48**:83-99.

Sands, D.C., Miller, R.V. and V.N. Hall (1982): In *Phytopathogenic Prokaryotes* Vol. I (Mount, M.S. and G.H. Lacyeds). Academic Press. pp. 65-75.

Schaad, N.W. (1982): In *Phytopathogenic Prokaryotes* Vol. I (Mount, M.S. and G.H. Lacy eds.). Academic Press. pp. 19-28.

Skerman, W.B.D., Mc Grown, V. and P.H. Sneath (1980): *Int. J. Syst. Bacteriol.* **30**:225-420.

Starr, M.P. (1981): In *the Prokaryotes: A Handbook of Habitats, Isolation and identifications of Bacteria* (Starr, M.P., Stolp, H., Truper, H.G., Balows, A. and H.G. Schlegal eds.) Springer Verlag. pp. 123-134.

Verma J.P., Singh, R.P. and H.D. Chowdhury (1980): In *Recent Advances in the Biology of Microorganisms* (Bilgrami, K.S. and K.M. Vyas eds.) Bishen Singh and Mahendra Pal Singh Publishers, Dehra Dun. pp. 95-106.

Vidaver, A.K. and M.P. Starr (1981): In *the Prokaryotes: A Handbook on Habitats, Isolation and Identification of Bacteria.* (Starr, M.P., Stlop, H., Truper, H.G., Balows, A. and H.G. Schlegal eds.) Springer Verlag. pp. 1879-1887.

Contemporary Approach to Palynological Science

P.K.K. Nair
Environment Resources
Research Centre,
P.B. 1230, Perookada,
Thriuvanthapuram -695 005.

INTRODUCTION

Pollen, being a unit in the partnership of reproduction, frui and seed production, has an immense role to play in plant improve ment. In addition to this primary role in reproductive biology an transmission of genetical changes and hereditary characters, th pollen has relevance in a variety of areas of human pursuits, healt and welfare. In fact, the contemporary approach in any scientifi discipline is that it ought to be people oriented, in terms of benefits However, there has been very little realization that the pollen const tute the resource of fundamental value affecting various aspects c human affairs although the records of its importance are found i literature since prehistoric times. The stone carving of the Assyria period during 700 BC depicted the pollination of datepalms b winged. Beings which amply shows that man understood the valu of artificial pollination in plant improvement (Maheshwari, 1945 Since the time the compound microscope has been invented, th structural aspects came to be understood and during the course c the last 100 years, the study and understanding of pollen expande in many directions with the result that the applications of the polle unit became clear and its ramifications are so complex that we d not have a complete understanding of the scenario even now However, it is fairly understood that its applications cover all aspect of science including Biophysics and even Mathematics and has th potential for new applications in several areas. Even the very funda nental aspect, pollen morphology, is becoming important in resolving the breeding system, of genetic resources, whether species, genus or ecosystems. In fact, the architectural diversity and the structural uniqueness are the two factors that have given pollen an importance in terms of its applications.

BASIC PRINCIPLES

The Angiosperms and its predecessor, the Gymnosperms, produce pollen in association with the origin and organisation of the flower morphosystem prior to which there were flowerless plants, in which the spores were the reproductive propagules. The spores and the pollen together, though have basic biological differences, have the common feature of a protective coat, and palynological science covers all facts of this reproductive phase in the life cycle of plants. In gaining a proper understanding of the palynological science,both pollen and the spores ought to be taken together in the genealogy of reproductive evolution as required. In fact, the very beginning of an organised life body has been in the form of a spore - like structure as discovered from the Swaziland geological system in Africa which during the course of evolution through a period of over 300 million years, diversified both in form and function, as required by the reproductive requirements of the respective geological periods. The haploid pollen unit, is the ultimate result of a long period of vegetative evolution of the gametophyte to suit the needs of reproduction with the origin and rise of flowering plants. In this process, there has been adaptational changes in the structural and functional behaviour of pollen, covering rather the entire reproductive machinery. It is already well documented that pollen dispersal mechanisms are anemophilous, entomophilous or hydrophilous confirming with each of which the structural differences has been also a natural phenomenon. For example, the anemophilous pollen being highly mobile, fall free, being nonsticky, and bestowed with simple ornamentation pattern among other structural features; in the case of entomophilous plants the pollen are sticky, heavily ornamented to suit their carrier, the insects. The hydrophilous types have a waxy coating and ornamented variously (Faegri and Iversen, 1974). In each case the pollen are distributed along unforeseen directions and places of sedimentation or accumulations at each of which they provide new materials for research enquiries. In this way, palynology is an environmental science with ample provision for not only a variety of transfer mechanisms but also for an enormous diversity of sculptural patterns, and metabolic behaviour.

The outer wall, the exine to be precise, embodies the taxonomic features serving the identifications of plants they belong to. Moreover, they are resistant to decay and destruction and therefore can stand the hazards of environmental situations they are carried to, during the course of their journey from the anther to the place of their final destination. It is these principles that made palynology a fruitful and stable science with potential use in many unpredictable areas of human affairs.

As mentioned earlier, man is at the centre of all benefits and the contemporary trends in palynology also ought to be looked at from that angle. Being a microscopic unit, the pollen has it own limitations by way of sustainable availability for exploitations in the way of other parts of the plant, like the leaves, bark, seeds or fruits. However, the practical use of the spores of *Lycopodium* is mentioned in pharmacopias, and so also of the pollen of *Cycas* (Chopra *et al.*, 1956) or *Typha* among a few other plants, in all of which the raw material is procurable in desirable quantities. By and large, the use of pollen in human affairs remain untapped, but with the growing progress in the study of cell biology and biotechnology, a greater possibility for utilization of pollen is emerging as evident from the fact that it is being used as vector for gene transfer technology in the area of biotechnology.

AREA OF CONTEMPORARY APPROACH

The current trend in research being production-oriented leading to human benefits, there has been a perceivable change in the direction of research studies and the applications with regard to **pollen**. The major areas of interest are agriculture and forestry, aerobiology and allergy, geoprospecting and palaeoenvironment including oil exploration, coal stratigraphy; apart from many other minor areas including the world-wide problem of biodiversity. Some of the major aspects of interest are analysed below.

Agriculture and Forestry

Pollen management has been one of the oldest as well as the youngest areas of inertest in view of its vital role in plant improvement and seed production among crop plant in particular and in other plants of economic importance in general. Among the food plants, the legumes and the grasses provide the staple food to more than half or almost all the people of the world and therfore there has been special attention given to their improvement. One of the most vital aspects of such improvement programmes is evidently the best exploitation of genetic variations (genetic resources) in which the main thrust has been the improvement in the technology of cryopreservations of pollen for use in breeding. However, there is immense awareness in recent years about the importance of the wild resources in evolving gene transfer, mechanisms for disease resistance along with pollen preservation of improved genotypes. With the increase in knowledge of DNA technology, genomic similarities of wheat, rice, maize and other grasses have been established with the result that the "primaeval grass genome" may in future find greater applications in the improvement of cereals and millets, in all of which the pollen is going to become a major system of hightech research, both basic and applied. In fact, pollen could emerge as the potential material for environmental monitoring including the effect of pollutants from any sources which possibly changes the entire genome system.

The story of haploid plants through pollen culture has already reached the stage of commercialisation in agriculture particularly with regard to rice. The "pollen cell" with its unique protective system is finding use in genetic transfer. With the improvement in biotechnology we are now in a situation of producing transgenic plants, again for improvement of plants, although there is a fear that the production of such transgenics may lead to genetic pollution in populations, with unpredictable hazards to the ecosystems and evolutionary biology. It is, therefore, evident that pollen biotechnology which can be traced back to the basic levels of artificial germinations, has gone far to the level of gene transfer technology, the results of which are going to be enormous and pollen is the most vital element in the whole system.

In the crops, the study of morphological evolution has been made possible with the development of ultrasturcture technology and this has been applied to resolve the breeding system of a few crop plants. In the case of *Arachis hypogea*, the ornamentation pattern could give sufficient identification on the genetic relationships amongst species (Nair *et al.,* 1990). With regard to rice the taxonomy is still far from the solution, and the small number of 24 species are resolved into 4 complexes - *Oryza ridleyi, O. meyerians, O. officinalis* and *O. sativa*, and the genetic resources are still under explorations and investigations. From the study of pollen of the cultivated plants (*O. sativa:* Asian; *O. grassilima*: African) the primary group could be identified to the *O. officinalis* complex by virtue of the occurrence of primitive elongate aperture forms as against the circular aperture system in many other *Oryza* complexes and also the grass family as a whole. Similar applications were also demonstrated with regard to *Gloriosa* (Ravikumar and Nair, 1986). Form studies of hybrids and their parents (eg.*Cajanus*) it has been shown that the hybridity is expressed at such ultrastructural levels as exine ornamentation (Ravikumar *et al.,* 1979) and all these examples are only demonstrative of the importance of palynology in the area of agriculture for plant improvement.

In the case of forestry also, all the above parameters can be applied and further there are ever so many potential areas of applications in which pollen of *Cannabis sativa* (a narcotic plant) in the air of coastal city of Trivandrum which could be used as an index for tracing unauthorised cultivation of the crops in the western ghat forests.

There are two areas, although traditional, which can be of importance in the study of biodiversity and population biology particularly of plants of economic importance and conservation value. For example air borne pollen or the insect bee-carried pollen will provide information, with a wide spectrum of applications. The pollen analysis of honey and honey bee loads has been suggested to be of potential value in planning for apiary gardens so as to provide the raw material required for bee survival and honey production throughout the year (Chaturvedi, 1973). All over the world the honey bees are the aim of improving production by increasing pollination efficiency. Bee carried pollen provides a reflection of bee forage, and such informations can be used with advantage in preparing maps on plant distributions in forest systems, and for naming honeys based on the dominant plant represented by pollen in the respective honeys.

The knowledge on airborne pollen and spores has multidirectional applications involving agriculture, industry, basic and experimental aspect studies and several other areas like forestry, health, energy, environment etc. with considerable socio-economic impact (Nair, 1994). In the area of agriculture, disease- forecasting through the survey of incidence of airborne spores, and the movement of the plant diseases are the most important in addition to which is the study of crop biodiversity in the context of the current trends in biological research. In the area of industry, the seasonal and geographical incidence of aerospora is of primary relevance in organising an antigen industry which is flourishing in the Western countries particularly. The medicinal importance of pollen and spores has a minor share of interest in pharmaceutical industry apart from the antigen factor. The knowledge from aerobiology has thus a wide spectrum of interest particularly because the spread of pollen and spores by air is the most potent in spreading the units to a variety of situations.

Oil Exploration and Coal Stratigraphy

Since the time Potonie (1956) formulated the principles and methods of pollen analysis, the "Sporae dispersae' have been used effectively in the correlations of coal seams and this traditional approach is still applicable. However, with the modern advances in computer technology, there is a new scope for wider and more efficient application of the information system on fossil spores and pollen. It should be noted that these fossils are the only unbroken record of plant evolution through geological time, right from the Precambrian to the Jurassic.

CONCLUDING REMARKS

The pollen spore morphology is the most unique structural system, in the spectrum of biology. The very fact that the exine is not only an embodiment of a combinations of characteristics of diagnostic value, but also is resistant to environmental hazards, during the course of sedimentation they become an index of ines-teemable value in a spectrum of areas through time in space. At the same time, as there was a structural evolution, there has been a corresponding evolutionary progress in the physiological phenomenon of the pollen spore units conforming with the pattern of evolution of the plants, particularly reflected in the reproductive system. As suggested by the enunciation of the triphyletic theory of the origin of

Angiosperms in the permocarboniferous, and has diverged along three directions, namely the `Magnolian' line, the Monocot' line and the `Ranalian' line at successive levels.

The above basic concepts are new to literature which might lead to far-reaching changes in the approaches to studies in plant systematics and phytogeography. A number of new areas are coming up among which mention may be made of the applications palynology in forensic science, live-stock ecology and even to test the power of blowers. The study of pleistocene vegetational history may be viewed from the point of view of gaining new knowledge on forest ecology for the planning of afforestation progarammes. It is imperative that the pollen and spores will emerge as an index of universal value in many areas of human pursuits and welfare.

ACKNOWLEDGEMENT

I am thankful to Dr. (Mrs.) S.K. Jayalakshmi, my scientific colleage of the Centre for helping in the preparation of the MS.

REFERENCES

1. Chaturvedi, M. (1973): *Grana* 13:139-355.

2. Chopra, R.N., Nayar, S.L. and I.C. Chopra (1956): *Glossary of Indian Medicinal Plants*, New Delhi.

3. Erdtman, G. (1969): *Handbook of Palynology*. Munksgaard, Copenhagan.

4. Faeqri, K. and J. Iversen (1974): *Text Book of pollen Analysis*. 3rd ed., Oxford.

5. Maheswari, P. (1945): *Nature* **3960**:354-355.

6. Nair, P.K.K. (1994): In *Recent Trends in Aerobiology, Allergy, and Immunology*. (ed. Agashe, S.N.) Oxford and IBH. New Delhi.

7. Nair, P.K.K. (1985): In *Proceedings of the International Symposium on Biological Monitoring of the state of the environment.* pp. 114-125.

8. Nair, P.K.K., Chaturvedi, M. and D. Yunus (1990): *Grana* **29**:109-117.

9. Potonie, R. (1956): *Beigh Goel* **23**:1-103.

10. Ravikumar, C., Nagaraj, M. and P.K.K. Nair (1979): *J. Palynol.* **15**:39-45.

11. Ravikumar, C. and P.K.K. Nair (1986): *New Botanist* **12**:1-68.

Ethnobotany of the Agastyar Hills in the Southern Western Ghats

P. Pushpangadan, S. Rajasekharan & K. Narayanan Nair

Tropical Botanic Garden and Research Institute, Palode, Thiruvanthapuram, 695 562.

INTRODUCTION

It is a well accepted fact that primitive societies like tribal and aboriginal communities are the repository of a unique treasure of knowledge base, accumulated over millenia, by their close associations, interaction and experimentations with surrounding environment. This rich and age old 'indigenous knowledge' or 'traditional ecological knowledge'(Gadgil *et al.,* 1993) embodies a great deal of information on indigenous people's perceptions on the natural environment and their traditional beliefs, practices and knowledge on the sustainable utilization of nature and natural resources, including biological diversity. Thanks to the emerging realization of the significance of such indigenous knowledge systems, especially their role in perfecting a holistic understanding of human ecology and natural resource management, concerted efforts have been intiated to study, document and conserve the biocultural diversity of the indigenous communities in several of the world's natural heritage sites, particularly in the tropics which represent the cradle of many an indigenous culture.

Ethnoecology is a broader scientific discipline which encompasses all studies which describe the indigenous and local people's interaction with the natural environment, including sub-disciplines such as ethnobiology, ethnobotany, ethnoentomology and ethnozoology (Martin, 1995). Ethnobotany thus forms a sub-discipline of ethnoecology which deals with the study of interrelationships between primitive men and plants (Jones, 1941; Plotkin, 1991).

The need for Ethnobotanical Investigations

In recent years, there has been a great surge of interest among scientists in pursuing ethnobotanical investigations, especially on the multifarious uses of plants as sources of food, medicine and all other necessities of life. This resurgence of interest has, by and large, emanated from the growing realization of the significance of ethnobotanical studies in finding new information on unexploited plant resources and new properties and uses of known plant resources as sources of food, medicine, fibre, fodder etc. This is becoming increasingly important in the realm of biodynamic plants which include medicinal, narcotic and toxic species (Schultes, 1991).

A recent estimation by WHO that '80% of the population in the developing countries depend on traditional medicine for their primary health-care needs substantiates the inevitable need and scope for revitalizing the traditional knowledge base ingrained in the rich and diverse cultural systems of tropical forest people. This is further augmented by the fact that out of the 120 active compounds that are currently isolated from higher plants and used in Western medicine, 74% have the same therapeutic use as in traditional societies. Information on plants used for medicinal purposes by primitive societies can provide us with new biodynamic compounds that may have significant applications in modern medicare systems. A classic example of this is the case of the rosy periwinkle, *Catharanthus roseus,* a plant native of Medagascar. The plant is a source of over 75 alkaloids, two of which *viz*. 'Vinblastine' and 'Vincristine' are now successfully used in the tratment of childhood leukemia and Hodgkin's disease. It is interesting to note that the discovery of these so called 'wonder drugs' was based on the primary information on the use of the rosy periwinkle as an oral hypoglycemic agent by the natives of Madagascar (Cordell & Farnsworth, 1976). The example of *Catharanthus roseus* and the like, therefore, invokes considerable insights into the prospect of undertaking chemical prospecting and pharmacological researches on tropical medicinal plants with a view to find out new kinds of biodynamic compounds that may lead to the formulation of new pharmaceutical drugs.

The private cultivars or wild plants used by tribal communities may hold key to many important agriculture breakthroughs by genetic engineering and other biotechnological innovations. Likewise, studies on the interaction of indigenous people with their ambient flora and fauna form a significant component linked with all planning and policy-decision processes related to the development of biosphere reserves, national parks and other protected area networks.

Recognizing the scope and utilities in tapping the unparalleled lore of traditional wisdom and knowledge, long term co-ordinated action oriented research programmes aiming at the survey, documentation and preservation of the biocultural diversity of the indigenous and local communities have been launched by several countries. The 'All India Co-ordinated Research Project on Ethnobiology (AICRPE)', launched by the Ministry of Environment and Forests, Government of India in 1982, is one such scheme which has already accomplished ethnobiological studies in about 85% of the tribal areas of India in two phases spanning from 1982 to 1988 and 1989 onwards. It is still an ongoing programme.

Ethnobotanical studies carried out by agencies under the AICRPE scheme have revealed that the various tribal communities in India use about 9500 species of plants for a variety purposes which include medicinal plants(7500), wild edibles (3900), other material and cultural requirements (700), fibre and cordage (525), fodder (400), pesticides, piscicides (300), gum, resin and dye (300) and incense and perfumes (100) (Anon.,1994).

Ethnobotany of the Agastyar Hills

The present account constitutes the results of ethnobotanical investigations carried out in the Agastyar hills of the Southern Western Ghats from 1987 onwards under the AICRPE scheme. While the major focus of the study was to survey, document and conserve the tribal/folklore medicine of Kerala, earnest attempts were made to document data on the other uses of plants too.

Study Site

Situated at the southernmost end of the Western Ghats, the Agastyar hills (Agastyamalai), the holy abode of the legendary sage Agastya, lies between 77°5' and 77°4'E and 8°20'and 8°50'N. Sprawled over an area of about 2000 sq.km. with altitude ranging from 67-1868 m, the hill range supports luxuriant vegetation and a much varied flora and fauna. The Agastyarkoodam which is the highest peak of the ranges is formed by a very compact block of hills comprising, Papanasam R.F., Singampatti R.F., Kakakkadu R.F., Kottur R.F., Kottur Extension R.F., Mahendragiri R.F., and Kalamalai R.F. (Henry *et al.,* 1982). The entire area ranges form part of Tirunelveli and Kanyakumari districts of Tamil Nadu and Trivandrum district of Kerala. The area receives an annual rainfall varying from 89 to 625 cm from two spells of monsoon rains : the South-West monsoon from June-September and North-East monsoon in October. Mean temperature varies between 21°C and 38°C with April and May forming the hottest and December to February the coldest months of the year.

Agastyar hills feature remarkable diversity in their ecosystems, flora and fauna. The vegetational diversity encompasses all representative kinds of forests in the western ghats such as the tropical thorn forests, dry deciduous forests, grasslands, tropical moist deciduous forests, tropical evergreen forests and subtropical montane forests. The flora of the region consitutes one of the richest and most diverse types with about 2000 species of flowering plants. Of these about 150 species are locally endemic to its different biomes. The important wildlife includes elephant, gaur, tiger, leopard, nilgiri lungur and lion-tailed macaque.

Tribal Settlements

'Kani' or 'Kanikkars', having the highest population among the forest dwelling tribals of Kerala, are the only tribe inhabiting in this area. The major tribal settlements covered in the present survey included Chinampara, Mangode, Podiyam, Pothode, Kaithode, Mulamoode, Ariyavila, Valipara, Amala, Plathus, Erumbyalu, Anakal, Aamode, Kunnatheri, Pattanpara, Cherumanga, Mlavila, Parandode and Kamalakam.

Kannis are of mixed racial origin having features of Dravidian, Veddoid and Negrito. They are traditionally nomads, once moved freely in the forest, subsisting on leaves, tubers, and fruits of wild plants and by hunting wild animals. They also practised shifting cultivation. Traditional occupation of Kani tribes include craft work like basket-making, mat-making using *Ochlandra* stem, cane works etc. They are also engaged in seasonal collection of minor forest produce like honey, bee wax, medicinal plants and python fat. The collective farming system once prevalent among the Kanis is now disappearing on account of their fast changing life styles and other socio-economic constraints. Currently they are engaged in the cultivation of edible plants such as tapioca, banana, millets and other cash crops like pepper, coconut, rubber, arecanut, cashew nut etc. They now represent one of the most civilized group of Indian tribes, though some are still leading a secluded life in the interior forest areas. Some of them are now employed in a hydro-electric project, private estates or the forest departments. A vivid account of their socio-economic setting, life styles, customs, beliefs etc. has been provided in the AICRPE Final Technical Report (1980-90), RRI (Drug Research), Poojappura, Thiruvananthapuram, and Shashi (1994).

Materials and Methods

Detailed survey, documentation, inventorty etc. were carried out in all the above-mentioned tribal settlements with the help and active co-operation from a good number of tribal extention officers, officers and staff of the Kerala Forest Department and elderly and knowledgeable tribal women and men. Door-to-door survey, interviews and documentation works were accomplished by methodical collection, cataloging, analysis and interpretation of data gathered on the very many uses of plants by the Kani tribals. Tribal claims on the medicinal and other uses of plants were subjected to further scrutiny, evaluation and standardization as per modern scientific methodologies. Vouchers of plant specimens collected were identified, processed and preserved in the herbarium (TBGT) at Tropical Botanic Garden and Research Institute. Live plants/seedlings were also introduced and nurtured in the medicinal plants garden at TBGRI.

Results

Extensive and intensive surveys and studies, resulted in the identification of over 300 species of higher plants used by the Kani tribe for a variety of their day-to-day requirements. Of these, about 150 species included wild edibles, economically important plants, fish poisons, fibre yielding, magico-religious, dye yielding and pesticides. The following enumeration includes the details of some of the less-known but most promising plants used by the Kanis.

numeration

Medicinal plants

1. *Anisomeles malabarica* R.Br. (Lamiaceae)

H : Undershrub up to 2m high.
VN : Perumthumba

Uses: Expressed juice of the fresh leaves (10-15 ml) taken internally.

2. *Aristolochia tagala* Cham. (Aristolochiaceae)

H : Twining herbs
VN : Valiya arayan, Mala arayan

Uses: Paste perepared from fresh or dry roots(5-10gm) administered internally along with human urine twice daily for snake posion, and also applied externally on the bitten part.

3. *Drynaria querciforlia* J. Smith (Polypodiaceae)

H : Stoloniferous herb; epiphytic
VN : Marappannakizhangu

Uses: Powdered mass of the stolons (10-15 gm fresh) along with previous day's rice gruel or with the juice of coconut kernal administered internally (thrice daily for 4 days) to cure jaundice, Salt free diet, and dip bath in the river recommended.

4. *Janakiya arayalpathra* Joseph et Chandrasekharan (Periplocaceae)

---

*H = Habit; VN = Vernacular Name (In Malayalam)

H : Erect shrubs, 1-2 m high
VN : Amruthapala

Uses: Expressed juice obtained from fresh tuberous roots (50 ml) mixed with equal quantity of coconut kernal juice is slightly boiled to prepare a semisoild mixture and it is taken internally twice daily for a period of 15 to 30 days for the cure of cancer like afflictions in the stomach, peptic ulcer etc. The mixture is also as a good blood purifier.

5. *Trichopus zeylanicus* Gaertn. sub sp. *travancorius* Burkill ex Narayanan (Trichopodaceae)

H : Herb, stem 2.5-12 cm long
VN : Arogyapacha

Uses: Daily intake of 4 or 5 tender fruits promotes stamina and vitality.

6. *Zornia diphylla* Pers. (Papilionaceae)

H : Prostrate herbs
VN : Muriva porinthi/Hemakandipaccha

Uses: Expressed juice of fresh leaves applied externally to check bleeding from cut wounds.

II. **Wild edibles**

7. *Aporosa lindleyana* (Wight) Baill. (Euphorbiaceae)

H : Small tree 8m tall
Vn : Vetti

Jses: Ripe fruits eaten.

3. *Arenga wightii* Griff. (Arecaceae)

H : Trees, main trunk 6-8m long
/N : Aayathengu

Jses: Toddy tapped from the inflorescence and peduncle used as refreshing drinks.

9. *Baccaurea courtallenis* (Wright) Muell. (Euphorbiaceae)

H : Trees, 10 m tall
Vn : Moottikai

Uses: Ripe fruits eaten.

0. *Dioscorea oppositifolia* Linn. (Dioscoreaceae)

: Climber, stem twinning to right
'N : Kavalan

ses: Root tubers eaten after boiling or in roasted form.

1. *Sterculia guttata* Roxb. *ex* DC. (Sterculiaceae)

: Trees, 10-15 m tall
N : Thondi

ses: Seeds eaten after frying.

III. **Economically Important Plants**

12. *Acacia caesia* (Linn.) Wild. (Mimosaceae)

H : Scandent shrubs
VN : Incha

Uses: Cleaning the body instead of using modern toilet, accessories

13. *Bambusa arundinacea* (Retz.) Roxb. (Poaceae)

H : Throny woody grass, 20-30 m tall
VN : Mula

Uses: Constructing huts, ladder and other artifacts.

14. *Calamus rotang* Linn. (Arecaceae)

H : Climbing palm
VN : Chooral

Uses: Making baskets, cot, chair etc.

15. *Dalbergia latifolia* Roxb. (Papilionaceae)

H : Trees, 15-20 m tall
VN : Eetti

Uses: Timber for constructing huts and making agricultural imp-ments.

16. *Ochlandra travencorica* (Bedd.) Benth ex. Gamble (Poaceae)

H : Woody grass, stems 2-6 m tall
VN : Era, Eetta

Uses: Stems used for making mat, basket and constructing huts.

IV. **Plants used for Fish Poisoning**

17. *Albizia chinensis* (Osbeck.) Merr. (Mimosacaee)

H : Trees, 15-24 m tall
VN : Peelivaka

Uses: Pounded mass of stem bark is dissolved in water for fish poisoning.

18. *Acacia concinna* DC. (Mimosaceae)

H : Small tree (cultivated)
VN : Neervalom

Uses: Seeds used for fish poisoning.

V. **Fibre yielding Plants**

19. *Helicteris isora* Linn. (Sterculiaceae)

H : Treelet, up to 5 m tall
VN : Idampiri valampiri

Uses: Fibre made out of the stem used for tying purposes.

20. *Agave americana* Linn. (Agavaceae)

H : A stemless plant with rossettee of leaves, native of America and West Indies

VN : Anakaitha

Uses: Fibre obtained from stem used for making nets and tying purposes.

Magico-religious plants

21. *Terminalia paniculata* Roth. (Combretaceae)

H : trees, 12-16 m tall
VN : Maruthu

Uses: Flowers offered in larger quantity before the hill diety during annual religious festival. The performance is called 'Poopada'.

22. *Poeciloneuron indicum* Bedd. (Clusiaceae)

H : Large trees, 20-30 m tall
VN : Poothamkolli

Uses: Sacred plant used for making spiritual sticks which are used in witchcrafts.

ACKNOWLEDGEMENTS

The work was carried out under the All India Co-ordinated Research project on Ethnobiology. Sponsored by the Ministry of Environment and Forests,Govt. of India, New Delhi.

REFERENCES

Anonymous (1994): *Ethnobiology in India - A Status Report,* AICRPE, Ministry of Environment and Forests, Govt. Of India, New Delhi.

Cordell, G. and N. Fransworth (1976). *Heterocycles* **4**(2):393-427.

Gadgi, M., Berkes, F. and C. Folk (1993). *Ambio* **22**:151-156.

Henry, A.N., Chanderbose, M., Swaminathan, M.S. and N.C. Nair (1994): *J. Bombay Nat. Hist. Soc.* **81**:282-290.

Jones, V. (1941): *Chron. Bot.* **6**:219-221.

Martin, G.J. (1995):*Ethnobotany, A methods Manual.* Chapman and Hall. London & Madras.

Plokin, M.J. (1991): *In the conservation of Medicinal Plants* (Akerele, O., Heywood, V. and Synge eds) Cambridge Univ. Press, Cambridge.

RRI (Drug Reaserch), Poojappura, Triuvananthappuram (1989-90): *AICRPE Final Technical Report.*

Schultes, E.R. (1991): In *The Conservation of Medicinal Plants* (Akerele, O., Heywood, V. and Synge eds). Cambridge Univ. press, Cambridge.

Shashi, S.S. (1994): *Encyclopedia of Indian Tribes.* Vol. **8**, Anmol Publications Pvt. Ltd., New Delhi.

Underutilized Food Legumes of India with Particular Reference to Himalayas

T. Husain and R.R. Rao
Taxonomy & Herbarium Section,
National Botanical Research Institute,
Lucknow-226 001.

The importance of legumes in human welfare is well known. Apart from the source of protein in our daily diet, legumes are utilized for a variety of fuel, timber and house hold materials, gums and resins, biochemicals and medicines and many other useful commercial products. Some are ornamentals or used for soil amelioration, or in environmental restoration (Anon. 1948-1976; Chopra *et al.*, 1956, 1969; Agarwal, 1986).

Although the family Leguminosae is richly represented in India, (Husain and Kapoor, 1990; Sanjappa, 1991) with an enormous biodiversity, barley 0.5% legume species are routinely consumed and are grown commercially. While these commercial crops cannot meet the protein demand of the enormous human population in the tropics, a large majority of the non-commercial food legumes go underutilized or even unnoticed. Some such legume species occurring in the wild or under localized cultivation in the Himalayan region are mentioned here for their popularization, commercial cultivation and utilization.

Non, Conventional Legumes of India

From time to time, botanists have highlighted the rich legume resources occurring in different biogeographic zones of India (Rao, 1958; Atal *et al.*, 1980; Thomas *et al.*, 1974; Gunjathkar and Vartak, 1982; Arora *et al.*, 1980). Much of the studies, however, pertain to the wild relatives of the established pulse crops.

These studies have confirmed that *Pueraria tuberosa* is an emergency food as a moderate source of calcium, iron and carbohydrates and also a good fodder. *Vigna angularis* (*Phaseolus angularis*) of Sino-Japanese origin, is a new promising pulse for the hilly regions of India. A survey in Pune yielded edible Legumes (Gunjatkar and

Vartak, 1982). *Vigna umbellata* is claimed to be a tribal pulse of eastern India: *Medicago sativa* (Lucerne), a forage plant, is recommended for mechanically separating its crude protein into a protein concentrate for use in human diets (Joshi, 1981). Garg and Oswal (1982) evaluated amino-acids of defeated seeds of *Caeslpinia sapan* for human consumption. Chaudhary and Singh (1886) reviewed potentials of 33 minor beans to supplement human diet.

The resurgence of interest in ethnobotanical studies during the last 2 decades or so, has brought to light numerous non-conventional wild legume species having food value (*Bauhinia vahlii, B. purpurea, B. variegata, Indigofera cassioides, I. dosua, Xylia xylocarpa, Cassia tora, C. obtusifolia, Eriosema himalaicium, Flemingia procumbens, Vigna umbellata, V. vexillata, V. unguiculata, Entada pursaetah, Mucuna gigantea, M. monosperma, M. pruriens, Parkia timoriaria, Glycine max, Eriosema himalacium, Desmodium umbellatum, Derris indica, Pithecellobium dulce, Pueraria tuberosa, Millettia extensa* etc.) Some of these 'crops' have already found their way to local tribal markets and village bazars. Yet there is no systematic survey of the diverse legume crops that are consumed by the tribal and rural populations in India.

Diversity in the Himalayan Food Legumes

The Himalayas are a store-house of plant wealth of all groups. On a recent study (Rao and Husain, 1983) the authors have established the enormous diversity of legumes in the Himalayas. Of the total 488 species belonging to 92 genera of legumes in the Himalayas, only 12-15 species are consumed by the local population in one form or the other. Although no systematic evaluation of the diversity in different groups is available, some of the widespread genera such as *Acacia, Albizia, Bauhinia, Caragana, Crotolaria, Desmodium, Indigofera* etc. exhibit great diversity, and some morphotypes in them are facing the danger of extinction. Therefore, the genera such as *Pueraria, Canavalia, Mucuna, Phaseolus, Medicago, Astragalus, Prosopis, Caragana, Trifolium, Lespedeza* etc. need to be studied in a priority basis for sustainable development. Some important wild species of food value in the Himalayan region are enumerated in Table 1.

Table 1: Enumeration of edible legumes of India (Himalayan ones are shown by an astrix)

Sl.NO.	Name	Parts Used
1.	*Acacia farnesiana*	Tender leaves
2.	*A. leucophloea*	Bark, Seeds and young Pods
3.	*A. nilotica* subsp. *indica*	Seeds
4.	*A. sinuata*	Leaves
5.	*Aeschynomene aspera*	Leaves
6.	**A. indica*	Leaves
7.	*Albizia myriophylla*	Fibrous bark
8.	*A. procera*	Pods
9.	**Atylosia scarabaeoides*	Seeds
10.	*Bauhinia malabarica*	Tender,Leaves,Seed and Buds
11.	*B. purpurea*	Leaves, Flower-bud, Pods
12.	*B. vahlii*	Seeds
13.	*B. variegata*	Flowers, Buds, Inflorescence
14.	**Butea monosperma*	Leaves
15.	*Caesalpinia pulchellum*	Petals
16.	*Canavalia ensiformis*	Green pods, Seeds
17.	*C. virosa*	Pods
18.	*Cassia floribunda*	Pods
19.	*C. nodosa*	Green pods
20.	*C. occidentalis*	Roasted seeds
21.	*C. tora*	Unripe fruits
22.	**Cicer microphyllum*	Fruits, Seeds
23.	**Cyamopsis tetragonoloba*	Pods
24.	**Desmodium triquetrum*	Leaves
25.	*Entada phaseoloides*	Seeds
26.	*E. pursaetha*	Seeds
27.	*Eriosema chinense*	Seeds
28.	**E. himalaica*	Tubers
29.	*Erythrina variegata*	Seeds
30.	*Flemingia variegata*	Seeds
31.	*F. vestita*	Tubers
32.	*Glycine max*	Seeds
33.	**G. soja*	Seeds
34.	*Indigofera caerulea*	Seeds
35.	**I. cassioides*	Flowers
36.	**I. cordifolia*	Seeds

37.	**I. dosua*	Flowers
38.	*Lablab purpureus*	Leaves,Flowers
39.	*Leucaena leucocephala*	Young leaves
40.	**Macrotyloma uniflorum*	Seeds
41.	**Medicago sativa*	Leaves
42.	*Mucuna cochinchinensis*	Young pods
43.	*M. deeringiana*	Seeds
44.	*M. gigantea*	Seeds
45.	**M. monosperma*	Seeds
46.	**M. pruriens*	Fruits, Seeds, young Pods
47.	*Neptunia oleracea*	As a pot herb (whole plant)
48.	*Pachyrrhizus erosus*	Fruits, Tubers
49.	*Parkia roxburghii*	Pods, Seeds
50.	*P. timoriana*	Pods
51.	*Phaseolus calcaratus*	Seeds
52.	*P. coccineus*	Tuberous roots
53.	*P. lunatus*	Seeds
54.	*P. vulgaris*	Leaves, Seeds
55.	*Pithecellobium dulce*	Aril surrounding the seeds
56.	*Prosopis chiliensis*	Ripe pods
57.	*P. cineraria*	Stem bark
58.	*Psophocarpus tetragonolobus*	Fruit
59.	*Pterocarpus indicum*	Fruits
60.	**Pueraria tuberosa*	Tubers, Leaves
61.	**Rothia indica*	Leaves
62.	**Sesbania bispinosa*	Seeds
63.	**S. cannabina*	Leaves
64.	*S. grandiflora*	Young flowers, Pods
65.	**S. sesban*	Seeds, Leaves
66.	**Smithia conferta*	Leaves
67.	**S. sensitiva*	Leaves
68.	*Tamarindus indicus*	Pods, young leaves
69.	**Trigonella corniculata*	Pods
70.	**Vicia faba*	Young pods
71.	**Vigna aconitfolia*	Immature pods, Seeds
72.	*V. adenantha*	Tubers, Seeds
73.	*V. mungo*	Seeds, Green pods
74.	*V. radiata*	Seeds
75.	**V. umbellata*	Seeds
76.	*V. unguiculata*	Young shoots, Seeds
77.	*Wisteria sinensis*	Seeds

The Grain Legumes: The Protein Resources of the Future

The grain legumes, as the word denotes, are important for their edible seeds. The Eastern Himalaya including the North eastern States are especially rich in the grain legumes (Kumar, 1988). Some of the important grain legumes of the region which have great potential as future protein sources are *Atylosia scrabaeoides, Canavalia ensiformis* (Jack bean), *C. gladiata* (Sword bean), *Glycine max* (Soyabean), *Lablab purpureus* (Hyacinth bean), *Mucuna nivea, M. utilis, Farkia javanica* (Tree bean). *Phaseolus vulgaris* (Kidney bean), *P. fuscus, P. lunatus* (Lima bean), *Psophocarpus tetragonolobus* (Winged bean), *Pueraria phaseoloides, Vicia faba* (Broad bean), *Vigna umbellata* (Rice bean) and *V. unguiculata* (Cowpea). Although *Psophocarpus tetragonolobus* is not native to India, maximum diversity noticed in this crop in North eastern states. The unripe seeds of this species contain about 35% of protein and 20% of edible oil. With detailed investigations on the extent of variations, followed by genetic improvement and development of suitable agrotechnology, there is no doubt that this crop can prove to be a wonder crop of the 21st century. Similarly, the two species of *Canavalia, Canavalia ensiformis* with white seeds and *C. gladiata* with broad red seeds and *Mucuna* with as many as six species in the north east India, have great potential.

Wild Relatives of Legume Crops

The diversity of legume-crops and that of their wild relatives occurring in India is of a high order. Major botanaical/phytogeographical areas of distribution of the latter, corresponding to areas with maximum species concentration are, the eastern and western peninsular tracts, the north-eastern parts and the western Himalaya (Arora, 1985; Arora and Nayar, 1984).

Flemingia vestita is a tuberous crop showing more variation of wild types in Western Himalaya (1800 m) but mainly domesticated in Khasi Hills (Meghalaya); it occurs in sub-Himalayan tract also, as forest undergrowth or in grassy patches.

India is a centre of diversity for Asiatic *Vigna* spp. Wild *V. mungo* (Khandala and Konkan belt in Western Ghats) is related to cultivated black gram. *Vigna khandalensis*, (*Phaseolus grandis*) has pods like those of *V. radiata, V. dalzelliana* (*P. dalzellii*) is related to the cultivated *V. angularis* and *V. umbellata* types (Peninsular hills, sub-Himalayan tract and north-eastern hills) are akin to their cultivated forms. Wild forms of *V. aconitifolia* (moth bean), occur sporadically in north and north-western plains and Deccan plateau; those of *V. trilobata* extending to sub-Himalayan tract. *V. vexillata* is another species occurring in sub-tropical Himalaya and peninsular hills, extending to north-eastern regions. *V. radiata* var. *sublobata* and *V. mungo* var. *sylvestris* are related morphologically and cyto-genetically to cultivated *V. radiata* and *V. mungo* (Jain and Mehra, 1980). *V. radiata* var. *sublobata* is widely distributed in Terai range, Siwalik hills, Kumaon upto 1200 m, Peninsular tract, Khandala ghats and Konkan southwards; the pods and seeds of this species are like those of *V. radiata* var. *setulosa*.

Concluding Remarks

Although the legume resources in this Indian region are quite rich and diverse, the protein deficiency in large majority of the population, especially rural and tribal folk, is quite evident. With the astonishingly high growth rate of population, particularly in tropical countries, the conventional legume resources certainly cannot meet the protein requirements of the masses, and added to this, there is always the danger of losing a crop variety due to pests and diseases. But at the same time, the very genetic estate of the country is at stake due to large scale habitat deterioration, particularly in Himalayas. Therefore, the need for survey and documentation of wild food resources for the 21st century has widened. Urgent studies are required to be undertaken on systematic assessment of genetic diversity and food value in all legumes. Similarly, agrotechnology for all grain legumes and above all, systematic inventorisation on the impact of these crops on the human system are other related aspects needing thorough investigation.

REFERENCES

Agarwal, V.S. (1986): *Economic plants of India*. Kailash Prakashan Calcutta.

Anonymous (1948-1976): The Wealth of India. Raw materials CSIR, New Delhi. Vols.1,XI.

Arora, R.K. (1985): *Genetic Resources of Less known cultivated Food Plants*. National Bureau of Plant Genetic Resources, New Delhi.

Arora, R.K. and E.R. Nair (1984): *Wild relatives of Crop Plants in India*. National Bureau of Plants Genetic Resources, New Delhi, India.

Atal, C.K., Sharma, B.M. and A.K. Bhatia (1980): *Indian Forester* **106**(3):211-219.

Chaudhary, B.D. and V.P. Singh (1986): *Haryana J. Hort. Soc.* **15**(1-2):133-146.

Chopra, R.N., Nayar, S.L. and I.C. Chopra (1956): *Glossary of Indian Medicinal Plants* CSIR, New Delhi, India.

Garg, S.C. and V.B. Oswas (1982): *Acta Siencia Indica* **8**(4):193-195.

Gunjatkar,N. and V.D. Vartak (1982): *J. Econ. Tax. Bot.* **3**(1):1-9.

Jain, H.K. and K.L. Mehra (1980): *Advances in Legume Science* **1**:459-468.

Joshi, R.N. (1981): *Indian J. Bot.* **4**(2):22-26.

Husain, T. and S.L. Kapoor (1990): *Enumeration of Legumes in India* (*Indigenous and Introduced*). National Botanical Research Institute, Lucknow, India.

Rao, P.S. (1958): *Indian Forester* **84**(5):281-83.

Rao, R.R. and T. Husain (1993): In *Himalayan Biodiversity Conservation Strategies*, (ed. U. Dhar, G.B. Pant) Inst. of Himalayan Environment and Development, Kosi, pp. 253-266.

Sanjappa, M. (1991): *Legumes of India*. Bishen Singh Mahendra Pal Singh, Dehra Dun.

Thomas, T.A., Patel, D.P. and N.R. Bhagat (1974): *Indian Fmg.* **23**(12):29-30.

Endangered Plants of the Himalayas and Their Conservation

R.R. Rao, B.Datt, and A. Prakash
Taxonomy & Herbarium Division,
National Botanical Reasearch Institute,
Lucknow-226001.

INTRODUCTION

The Himalayas are the world's highest mountain ranges extending east to west from the river Indus to river Brahmaputra over 2400 km. with an estimated width varying between 160 to 400 km. They are divided into three parallel longitudinal zones namely (i) the Great Himalaya or Trans-Himalaya in the north with average elevation of 6000 m. (ii) the Lesser Himalaya or the Middle Himalaya with average elevation of 4500 m, and (iii) the Outer Himalaya or Siwalik range from Gangetic Plain to 1200 m high. Vertically these ranges show a wide altitudinal range from tropical foot hills or the outer Himalaya to the peaks as high as 8848 m in Mount Everest. Biogeographically the Indian Himalayas can be classified into three bioprovinces, namely (i) The Trans-Himalaya forming the 'Cold desert' of the India comprising of Ladakh district of Jammu and Kashmir State and Lahaul-Spiti district of Himachal Pradesh, (ii) Western Himalaya comprising of Jammu and Kashmir, Himachal Pradesh and Kumaun and Garhwal regions of Uttar Pradesh and (iii) Eastern Himalaya comprising of Arunachal Pradesh, Sikkim and Darjeeling districts of West Bengal. The Himalayas have a diverse climate resulting in cold arid desert of Ladakh in the north-western region to the lush green tropical and temperate forest of Arunachal Pradesh in the east. With the background of most outstanding and complex topographic features, it is not surprising that the Himalayas have a vegetation far richer and more varied than that of many parts of the subcontinent, if not the globe. Moreover, the Himalayan region serves as the store-house of numerous highly valuable medicinal plants, rare horticultural and oranmental plants, many botanical curiosities and also species which are known to be endangered or threatened. For the past few decades the plant resources of India including the Himalayas are under tremendous stress on account of several factors. The problem of destruction of habitat in the Himalayas, and of endangered species has today acquired such prominence that it has accumulated considerable literature (Santapau, 1971; Sahni, 1973, 1980, Raizada, 1977, 1980; Jain, 1976; Jain & Sastry, 1980; Maheshwari, 1980, 1984; Nayar; 1977; Rau, 1981; Jain & Rao, 1983; Nayar & Sastry, 1987-90; Rao, 1994). Yet the endangered species are poorly understood.

Endangered species

Endangered species are those whose reproductive capacity is far lower than the number of plants estimated from the habitat due to natural or anthropogenic factors. This consequently leads to the decline in number and size of populations. Conservation of such endangered species, though sounds meaningful and acceptable to all conservation biologists, is not an easy task, nor it could be achieved just by declaring certain areas in the Himalayas protected or shifting a few plants to the botanic gardens for *ex situ* conservation. Although species disappearance attracts wide publicity, the loss of genetic diversity due to loss of population in a species is little appreciated nor there are any case-studies to support such a conclusion. Extensive field surveys are the foundation for such studies and these should be encouraged. Also the correct taxonomic concept is very much needed for proper identification of endangered species. Certain species reported to be rare in the Himalayas are shown to be abundant by subsequent workers and this is mainly due to wrong identification.

ow to decide a species endangered ?

To decide a species as endangered or threatened is not an asy task but requires very elaborate field assessments. Customarily a species is not represented by more collections from different calities in a herbarium, such a species is determined as a rare pecies and this may not always be correct. Sometimes if a species collected after a gap of 25 to 50 years then also it is concluded as n endangered species. While partly it could be true, certain ephimral plants such as *Thermopsis barbata* and some species of *Aconiım, Gentiana* and *Primula* which complete their life cycle in a short eriod and die off are likely to be missed by botanists visiting these reas. Hence not collecting a plant for several decades may not ecessarily conclude a species as endangered.

All the endangered plants are certainly rare but not all rare lants are endangered. For e.g., several sparsely distributed alpine lants like few species of *Saussurea, Impatiens, Primula, Corydalis, 'halictrum, Saxifraga* etc. have developed ecological or reproductive daptations to persistence in small populations showing natural rarity nd are genetically stable. On the other end of the spectrum there re some rare species, which were supposed to be widespread at ne time, have now undergone a rapid decline due to anthropogenic actors (anthropogenic rarity). For e.g. *Dioscorea deltoides, Podophylım hexandrum, Aconitum heterophyllum, Nardostachys grandiflora, llium stracheyi, Panax pseudoginseng, Picrorhiza kurrooa, Angelica lauca* etc. These aspects require very extreme and first hand field ıvestigation before one can decide a species endangered or not. ccording to the minimum viable population concept (Shaffer, 1981) minimum of 50 individuals are recommended for maintenance of he population. Those species with less than 50 individuals in a opulation are certainly rare.

Causes of threat

Due to natural and man-made (anthropogenic) factors, th plant resources of the Himalayas are under varying degrees of threa and such a threat is rather accelerating over the years. It is now we known that several species have become endangered due to certai natural phenomena, such as land sliding, glaciation and forest fires In the recent years, developmental activities such as construction c dams and roads, extensions of townships, mining operations and als encroachment of vast forest areas for shifting cultivation (in easter sector), are examples of man-made imbalances in the ecosystem. T these can be added the large scale timber extraction from forests. large number of orchids, therefore, are under tremendous stress du to removal of the habitats. Besides, the selective but large scal removal of certain medicinal plants such as *Podophyllum hexar drum, Coptis teeta, Dioscorea* spp.,*Allium stracheyi, Nardostchy grandiflora, Rheum emodi, Aconitum* spp., *Picrorhiza kurrooa* an *Saussurea lappa,* has also added to the depletion of more importar wild economic plants. It is no wonder that many of these valuabl plants are not found in large populations in their natural habitats Steps should therefore, be taken to ensure that plant diversity conserved. Apart from the above threats, the introduction and sprea of alien weeds such as *Mikania scandens, Chromlaena odorata* an *Parthenium hysterophorus* have greatly altered the original ecosys tems. Several native species have been completely eliminated fro the area. Once the species gets fragmented and the demography c the population is altered, the genetic diversity that is essential fc sustaining the species also gets eroded.

Conservation

Although a few puplications on endangered species of tl Himalayas have emerged during the last one or two decades, thes uplications simply deal with the species as a whole and no efforts re made to evaluate the number and size of population of an endan- ered species, the actual causes of threat, etc. This later part re- uires rather very extensive and frequent field surveys streching the ntire distribution range of a particular species, though it is certainly xpensive, time consuming and difficult. A large number of species o far listed as endangered in the Himalayas are based, in the majori- y of cases on the scrutiny of herbaria. Therefore, what is now re- uired is the assessment of individual species in its habitat itself, so hat the correct status of the endangered species can be determined.

Genetics and demography *i.e.*, population dynamics, are rucial to success in long term conservation management of any pecies. For e.g., in case of *Eremostachys superba*, the population f which is small and isolated near Mohund in the Siwalik of Uttar radesh, a strong genetic drift operates. Population in which drift perates lose variation more readily than do population in which drift not a factor. Another consequence of low population structure is ne occurrence of inbreeding resulting in loss of fitness (inbreeding epression). Therefore, we have to think of strategies that increase ne genetic diversity in rare species by bringing together geographi- ally isolated populations. This again requires extensive survey to nd out the distribution of various populations of a given endangered pecies.

The reproductive biology and breeding system of rare species another aspect of study which is almost neglected. The ongoing productive biology studies on *Eremostachys superba* at NBRI, ucknow, has shown that (a) its population has drastically collapsed hardly 25 plants at Mohand (b) the species is adapted for out- reeding but forced to inbreed, (c) refuse to inbreed but when polli- ated show high seed set (44.4%),(d) low plant population affected ne pollinator population, (e) seed germination failed under natural conditions and, (f) poor sexual reproduction and failure of see formation has resulted in development of the thick rootstock syster to serve as buffer from environmental stochasticity. From the stud it is clear that preservation of the habitat alone, as attempted i most cases, may not assure the longer term conservation of thi species. Further, the loss of diversity, both phenetic and geneti reduces the ability of populations to adapt to changing environmen and increase their susceptibility to pests and diseases. These plan need to be protected on top priority. In a situation where sever endangered species grow, *e.g.* within a few hectare or sq. kms., th whole area should be protected. The occurrence of several endem species in a particular area also indicates the potential of that are as a centre of speciation or evolution. Secondly when an endangere species grows in pure population or a few or more individuals, or i association with other vegetation, it becomes necessary to establis a sanctuary for the individual plants e.g., the *Rhododentron* sanctu ary in Sikkim. In addition, *ex situ* conservation in botanic garden seed banks or *in-vitro* cultures of endangered species should also b attemped side by side.

Once the status of endangered species of the Himalayas determined, the next step would be to bring out some sort of public ty material along with pictures of the plants and their conservatio significance preferably in different local languages understood by th people in the Himalayas. The local communities should be involved every stage for conserving a species in its habitat.

Monitoring of population of the endangered species is anot er area of work which is not attemped in the Himalayan regio Monitoring of the species involves the study of the populatio dynamics during different seasons over a period of time. This giv us an idea about the adaptive ability of a particular endanger species to its natural habitat. Where required, overgrowth of oth surrounding species also need to be controlled to keep the populati of the endangered species to a level it can sustain itself.

The rare and endangered plants of the Himalayas are listed in table I based on data available in flora, relevant literature, herbaria as well as field observations. Out of 188 endangered species in the Himalayas, 105 species are confined to temperate zone, 50 species in subalpine and alpine zones while 33 species extend their distribution from temperate to alpine zone in the Himalayas. It is observed that most of endangered species (90%) are herbs or undershrubs. Among the economically useful species, most of the species are of medicinal value. As the Himalayas are the only habitat for these threatened species and several of them have already declined in their population, it is necessary to conserve all the taxa by developing a network of protected areas and sanctuaries at suitable places as has done by declaring Namdapha (E. Himalaya) and Valley of Flowers and Nandadevi (W. Himalaya), as biosphere reserves.

To conclude, conservation of endangered species is a great challenge to all conservation biologists of the 21st century. The effort should not end in just listing certain species as endangered or declaring certain areas protected for conservation of certain rare populations. Even, if attemted so, the populations are threatened by obnoxious weeds, herbivores or demograhic or environmental stochasticity. Co-ordinated multidisciplinary approaches to work out strategies that increase the diversity in them is highly warranted.

ACKOWLEDGEMENTS

The authors are thankful to the Director, NBRI, Lucknow, for facilities.

REFERENCES

Chatterjee. D.(1940): *J. Asiat. Beng.* (N.S.). **5**:19-67.

Jain, S.K. (1976): *Indian Sci. Ann.* **2**:20-23.

Table I A list of the rare and endangered taxa of the Himalayas

Name of species	Distribution	Status
(1)	(2)	(3)
Ranunculaceae		
1) *Aconitum atrox*	Subalpine to alpine regions, 3300-3900 m.	Medicinal, rare
2) *A. deinorrhizum*	J.K., H.P., Nepal, Bhutan, 3000-4500 m	Endemic, vulnerable due to over-exploitation for medicinal roots
3) *A. falconeri* var. *latilobum*	Alpine regions	Vulnerable due to exploitation
4) *A. ferox*	H.P. to Sikkim, 3300-5000 m	Vulnerable due to overexploitation
5) *A. heterophyllum*	J.K.,H.P.,Kum., Garh., Nepal, 2000-3300 m	Medicinal, rare
6) *A. laeve*	J.K.,H.P.,Kum., Garh., Nepal, 2000-3300 m	Medicinal, rare
7) *Coptis teeta*	Arunachal Pradesh, 2500-3000 m	Endemic, vulnerable due to over-exploitation for medicinal rhizome
8) *Delphinium uncinatum*	J.K.,H.P., Pakistan, Afghanistan	Vulnerable due to habitat loss

	(1)	(2)	(3)
9)	*D. vestitum*	J.K. to E. Nepal 2500-4000 m	Medicinal, rare
10)	*Thalictrum alpinum*	J.K. to Bhutan, 3000-4000 m	Medicinal, rare
11)	*T. chelidonii*	J.K. to Kum., Sikkim, Nepal, Myanmar, 2500-3000 m	
Berberidaceae			
12)	*Berberis affinis*	Kum., Garh., 2300-2400 m	Rare,endemic
13)	*B. apiculata*	H.P., 3600-4200 m	Endemic, rare
14)	*B. huegeliana*	J.K.	Endemic, indeterminate
15)	*B. kashmirana*	J.K., 3300-3500 m	Rare, endemic
16)	*B. koehneana*	Kum., Garh., 2400-3000 m	Medicinal
17)	*B. lambertii*	Kum., Garh., up to 3000 m	Endemic, known by a solitary collection
18)	*B. osmastonii*	Garh. upto 3000 m	Endemic, rare, known from a solitary collection
19)	*B. petiolaris* var. *garhwalana*	Kum., Garh., 2400-2500 m	Endemic
20)	*B. royleana*	J.K., 2400 m	Endemic
Podophyllaceae			
21)	*Podophyllum hexandrum*	J.K. to Arunachal Pradesh 2400-4500 m	Medicinal, endangered

	(1)	(2)	(3)
Papaveraceae			
22)	*Dicranostigma lactucoides*	Kum., Nepal, 3200 m	Endemic
23)	*Meconopsis latifolia*	J.K., 2700-4600 m	Endemic, vulnerable due to natural causes
24)	*Papaver himalaicum*	J.K., 3600 m	Endemic
Brassicaceae			
25)	*Arabis tenuirostris*	J.K., 4500 m	Endemic
26)	*Draba dasyastra*	J.K., 4000 m	Endemic
27)	*Erophila tenerrima*	J.K. (Ladakh), 4200 m	Endemic, known from type only
28)	*Erysimum thomsonii*	H.P.	Extremely rare, known from type only
29)	*Thlaspi andersonii*	J.K. to S.E. Tibet 3000-4400 m	Endemic
Capparaceae			
30)	*Capparis pachyphylla*	Arunachal Pradesh	Vulnerable, endemic
Violaceae			
31)	*Viola himalayensis*	J.K., 4000 m	Endemic
32)	*V. kumawarensis*	J.K. to C. Nepal, 3000-4700 m	

	(1)	(2)	(3)
Caryophyllaceae			
33)	*Arenaria curvifolia*	Garh., 3350-3650 m	Endangered, not collected after 1885
34)	*A. ferruginea*	Kum.	Endemic, not collected after 1886
35)	*Cerastium thomsoni*	W. Himalaya, 3000-3600 m	Endemic
36)	*Dianthus minimus*	J.K., 4200 m	Endemic
37)	*Silene kumaonensis*	Kum., Garh., 2500-4000 m	Endemic, rare
38)	*S. kunawarensis*	J.K.,H.P., 2300-3300 m	Rare, known from very few collections
39)	*Stellaria depressa*	J.K., 4000 m	Endemic
40)	*Thylacospermum caespitosum*	J.K. to Sikkim, Tibet, C. Asia, 4500-5700 m	Monotypic
Balsaminaceae			
41)	*Impatiens jaeschkei*	Kum., 2700-3000 m	Ornamental, rare
42)	*I. harrissii*	W. Himalaya, upto 3200 m	
43)	*I. langeana*	Kum., upto 3300 m	Ornamental, rare
44)	*I. meeboldii*	J.K.	
45)	*I. pahalgamensis*	J.K., 3000 m	Endemic
46)	*I. podocarpa*	Garh., 2500 m	
47)	*I. polysciadia*	Kum., upto 3000 m	

(1)		(2)	(3)
48)	*I. reidii*	Kum., upto 2800 m	Ornamental, rare
49)	*I. stoliczkai*	H.P., (Kinnaur), 3000 m	
Celastraceae			
50)	*Euonymus fortunei*	Arunachal Pradesh	
Vitaceae			
51)	*Cissus spectabilis*	Sikkim, Darjeeling	Endemic
Fabaceae			
52)	*Astragalus bakeri*	J.K., 3600 m	Endemic
53)	*Hedysarum astragaloides*	J.K., H.P., 3500-4500 m	Endemic, rare
54)	*H. cachemirianum*	H.P., 3700-4000 m	Endemic, rare due to habitat loss
55)	*H. microcalyx*	J.K., H.P., Garh., 3000-4800 m	Vulnerable due to destruction of habitat
56)	*Indigofera simlensis*	H.P.	Endemic
57)	*Pueraria bella*	E. Himalaya, N.E. India	Rare
58)	*Thermopsis inflata*	J.K., H.P. (Ladakh), H.P. (Lahul), 4000-5000 m	Endemic
59)	*Vicia benthamiana*	J.K.	Endemic

	(1)	(2)	(3)
Rosaceae			
60)	*Cotoneaster cashmirensis*	J.K., 3300 m	Endemic, ornamental
61)	*C. simonsii*	Sikkim, 1545-3152 m	Endemic, indeterminate, not collected after 1904
Begoniaceae			
62)	*Begonia arborensis*	Arunachal Pradesh	Endemic, rare
63)	*B. burkillii*	Arunachal Pradesh	Endemic, rare, known from type only
64)	*B. rubella*	Sikkim,Nepal upto 1800 m	Rare, known from type only
65)	*B. satrapsis*	Sikkim,Darjeeling	Rare
66)	*B. scintillans*	Arunachal Pradesh	Endemic, known from type only
67)	*B. scutata*	Sikkim,Darjeeling	Rare
Apiaceae			
68)	*Acronema pseudotenera*	N. Sikkim, 3700 m	Indeterminate, not collected since 1892
69)	*Angelica nubigena*	N.E. Sikkim, 3800 m	Endemic, known from type only
70)	*A. glauca*	W. Himal. 2100-3300 m	
71)	*Chaerophyllum cachemiricum*	J.K., H.P.	Endemic, known from type only

	(1)	(2)	(3)
72)	*Heracleum jacquemontii*	N.W. Himalaya upto 3600 m	Endemic, known from type only
73)	*H. thomsonii*	J.K. > (Ladakh), H.P. (Lahul) upto 4000 m	
74)	*Ligusticum marginatum*	J.K., H.P., 2000-3000 m	
75)	*Peucedanum thomsonii*	J.K. (Kishtwar and Banahal) 1500-3000 m	Endemic
76)	*Pimpinella tongloensis*	Singaleela range in Darjeeling, Sikkim	Endemic
77)	*Pteropetalum radiatum*	N.Sikkim (Yumtang and Sebul Valley), *ca* 3500 m	Endemic
Araliaceae			
78)	*Panax pseudo-ginseng*	E. Himalaya, C.Nepal to S.W. China, N. Myanmar, 2900-4000 m	Rare
Valerianaceae			
79)	*Nardostachys grandiflora*	Kum., Garh., Nepal, Bhutan, Tibet, W.China, 3000-5000 m	Vulnerable due to over-exploitation of medicinal root-stock

	(1)	(2)	(3)
Rubiaceae			
80)	*Hedyotis scabra*	E.Himal., Bangladesh, Myanmar	Rare
81)	*Rubia edgeworthii*	N. Himal.	Vulnerable due to habitat loss
82)	*R. himalayensis*	J.K.,Pakistan, Afghanistan	Vulnerable
Asteraceae			
83)	*Catamixis baccharoides*	Garh., Nepal	Endemic, Vulnerable
84)	*Cremanthodium plantagineum* f. *ellissii*	J.K., H.P.	Rare
85)	*Chrondrilla setulosa*	J.K., upto 2800 m	Known from type only
86)	*Inula racemosa*	J.K., 3000-3500 m	
87)	*Lactuca benthamii*	J.K., 4000-5000 m	Endemic, known from type only
88)	*L. cooperi*	Sikkim, *ca* 5000 m	Endemic, known from type only
89)	*L. filicina*	Kum., (Kali valley), 1820-2430 m	Endemic, known from type only
90)	*L. undulata*	J.K.	
91)	*Saussurea bracteata*	J.K., H.P., Kum., Garh., 3500-5500 m	Rare
92)	*S. clarkei*	J.K., 4000-4500 m	Endemic, rare due to overgrazing
93)	*S. costus*	J.K.,H.P.,Kum., Garh.,3200-3800 m	Endangered due to over-exploitation for its medicinal roots

	(1)	(2)	(3)
94)	*S. obvallata*	J.K. to S.W. China	Medicinal, religious value
95)	*Senecio mishmi*	E. Himalaya	Indeterminate, known from type only
Campanulaceae			
96)	*Campanula wattiana*	H.P., U.P., 2200-3900 m	Endemic, rare
97)	*Codonopsis affinis*	Sikkim, Darjeeling	Endemic, rare
98)	*Cyananthus integra*	Kum.,Garh.	Endemic, rare
Primulaceae			
99)	*Primula reiddii*	J.K. to C.Nepal, 3300-4500 m	Rare
Styraceae			
100)	*Alniphyllum fortunei*	Arunachal Pradesh, China	Rare due to habitat loss
101)	*Huodendron biaristatum*	Arunachal Pradesh, China, Myanmar	Rare
Asclepiadaceae			
102)	*Ceropegia hookeri*	Sikkim, Nepal, Tibet	Known by only two collections in Indian herbaria

(1)		(2)	(3)
Gentianaceae			
103)	*Gentiana inflexa*	H.P., Kum., Garh., Sikkim, Tibet, China, 2775-4901 m	
104)	*G. sajinoides*	Kum., 4000-5000 m	Endemic, known from type only
105)	*G. tetrasepala*	Garh. (Ralam Valley), 4000-5000 m	Endemic, known from type only
106)	*Lomatogonium brachyantherum*	J.K.(Ladakh), H.P. (Spiti), Tibet, 3550-4600 m	
107)	*Lomatogonium brachyantherum*	J.K. (Ladakh), H.P. (Spiti), Tibet, 3350-4000 m	
Boraginaceae			
108)	*Arnebia benthamii*	J.K.,H.P.,Kum.,Garh., 3000-4000 m	Medicinal, rare
Solanaceae			
109)	*Pauia belladona*	Arunachal Pradesh	Rare due to habitat loss
Ericaceae			
110)	*Rhododendron concinnoides*	Arunachal Pradesh	Endemic, ornamental

(1)		(2)	(3)
111)	*R. falconeri* ssp. *eximium*	Arunachal Pradesh, 2400-3300 m	Endemic
112)	*R. nuttalli*	Arunachal Pradesh to China, 1200-3650 m	Ornamental, rare
113)	*R. santapauii*	Arunachal Pradesh, upto 2300 m	Endemic
114)	*R. subansiriense*	Arunachal Pradesh, 2600-2800 m	Ornamental, rare
Scrophulariaceae			
115)	*Picrorhiza kurrooa*	J.K. to Sikkim, 3300-5000 m	Vulnerable due to over-exploitation for its medicinal roots
116)	*Wulfenia himalaica*	Kum., *ca* 3000 m	Endemic
Circaesteraceae			
117)	*Circaester agrestis*	Kum., Garh., Nepal, Bhutan, S.E. Tibet, China	Rare
Balanophoraceae			
118)	*Balanophora involucrata*	J.K. to Sikkim, W. China, 2000-3000 m	Medicinal, a curious root parasite

(1)	(2)	(3)
119) *Rhopalocnemis phalloides*	E. India, E..Himalaya	A rare root parasite
Rafflesiaceae		
120) *Sapria himalayana*	Arunachal Pradesh, Manipur, Meghalaya	Rare due to destruction of habitats, surviving only in Namdapha area.
Gesneriaceae		
121) *Rhynchoglossum lazulinum*	Arunachal Pradesh	Endemic, rare
Lamiaceae		
122) *Eremostachys superba*	J.K.,H.P.,Garh., Pakistan	Highly endangered
Polygonaceae		
123) *Rheum emodi*	W.&C. Himal.	
Orchidaceae		
124) *Aphyllorchis gollani*	Restricted to Tehri Garhwal, 2400 -3000 m	Saprophytic orchid possibly extinct, not collected after 1900

	(1)	(2)	(3)
125)	*A. parviflorus*	Garh., Sikkim (Lachen and Yumthang), 3600-3700 m	
126)	*Archineottia microglottis*	Garh., 1800-2000 m	Endemic, rare
127)	*Bulleyia yunnanensis*	Arunachal Pradesh, Darjeeling, Bhutan, China, *ca* 2000 m	Rare due to deforestation
128)	*Calanthe alpina*	Kum., Garh., Sikkim, Nepal, Bhutan, 2500-3500 m	Rare
129)	*C. mannii*	E.Himal., Kum., Garh.	Rare, known from a few localities
130)	*C. pachystalix*	H.P., Garh., Nepal, Indo-China, *ca*. 2000 m	Known from its type specimens only
131)	*C. plantaginea*	J.K., H.P., Kum., Garh., Nepal, Burma, upto 2800 m	Rare
132)	*Coelogyne treutleri*	Sikkim	Possibly extinct. Known by type collection
133)	*Corallorhiza trifida*	J.K., Kum. to Nepal, *ca* 3500 m	
134)	*Cymbidium eburnum*	Arunachal Pradesh, N.E. India	Vulnerable due to loss of habitat
135)	*C. hookerianum*	Kum., Sikkim, Arunachal Pradesh, Nepal, Bhutan	Vulnerable
136)	*C. whiteae*	Sikkim, 1500-1700 m	Endemic

	(1)	(2)	(3)
137)	*Cypripedium elegans*	Garh. to Sikkim, 2500-4000 m	Rare due to loss of habitat
138)	*C. cordigerum*	J.K., H.P., Garh., Nepal, Bhutan and Pakistan 2500-3000 m	Rare due to loss of habitat
139)	*C. himalaicum*	Garh., Kum., Sikkim, Nepal, Bhutan, S.E. Tibet, 3000-4300 m	Rare
140)	*Dactylorhiza hatagirea*	J.K.,H.P.,Kum.,Garh., 3000-4000 m	Medicinal
141)	*Didiciea cunninghamii*	Garh., Sikkim, *ca* 4000 m	Monotypic
142)	*Diplomeris hirsuta*	Kum., Arunachal Pradesh, Nepal, Bhutan	Vulnerable
143)	*Eria occidentalis*	Kum., 1550 m	Rare, known from type only
144)	*Flickingeria hesperis*	Kum.	Endemic
145)	*Herminium duthei*	Kum.,Garh., 2800-4000 m	Rare
146)	*Listera pinctorum*	Garh., to Arunachal Pradesh, 3000-3500 m	
147)	*L. tenuis*	Garh., Arunachal Pradesh, Nepal, 3000-4000 m	
148)	*Neottia inayatti*	J.K., 3000 m	Rare, known from a few old collections only
149)	*Oreorchis indica*	H.P., Kum., Garh., 2000-2500 m	

	(1)	(2)	(3)
150)	*Paphiopedilum fairieanum*	Arunachal Pradesh, Sikkim, Bhutan, 1400-2200 m	Endemic due to anthropogenic causes
151)	*P. venustuma*	Meghalaya, Sikkim, Bangladesh	Vulnerable due to loss of habitat
152)	*P. wardii*	Arunachal Pradesh	Endemic, known from type only
153)	*Pholidota wattii*	Arunachal Pradesh, Assam	Endemic, rare
154)	*Zeuxine pulchra*	Sikkim, Meghalaya, *ca* 2000 m	Collected after 80 years; highly endangered
Zingiberaceae			
155)	*Roscoea alpina*	J.K. to Nepal, 2500-3600 m	Endangered, known from type only
Dioscoreaceae			
156)	*Dioscorea deltoidea*	J.K. to Assam, Afghanistan, Indo-China, W. China, upto 3500 m	Rare
Iridaceae			
157)	*Iris duthei*	Kum., up to 3600 m	Endangered
158)	*I. milesii*	J.K. to U.P. upto 2700 m	Rare

	(1)	(2)	(3)
Liliaceae			
159)	*Allium auriculatum*	H.P., J.K.	Endemic
160)	*A. loratum*	J.K., 3300-4200 m	Endemic
161)	*A. stracheyi*	J.K., H.P., Kum., Garh., Nepal 2400-2700 m	Vulnerable due to overgrazing
162)	*Dipcadi reidii*	W. Himalaya, *ca* 2300 m	Presumed extinct, known only from a single gathering 100 years ago
163)	*Eremurus himalaicus*	J.K.,H.P.,U.P., 2400-3300 m	Rare due to destruction of its habitat
164)	*Lilium podophyllum*	J.K.,H.P.,U.P., Pakistan,Afghanistan, 2100-3000 m	Ornamental, rare
165)	*Lloydia himalensis*	J.K.,H.P.,Sikkim, 3695-3810 m	Rare
166)	*Nomocharis synaptica*	Arunachal Pradesh (Thanu Valley, Delleri Valley), Myanmar, 3048-3695 m	Rare, so far collected from two localities
Juncaceae			
167)	*Juncus sikkimensis*	Sikkim, 4000-4500 m	Rare

(1)		(2)	(3)
Arecaceae			
168)	*Calamus inermis*	Sikkim	Endemic
169)	*Livistonia jenkinsiana*	E.Himal.,N.E.India	Endangered due to deforestation
170)	*Phoenix rupicola*	E.Himal.,N.E.India	Rare to to over-exploitation of edible pith
171)	*Trachycarpus takil*	Kum.,(Thalkeder hill, Pith oragarh Dist.) 2000-2500 m	Endemic, rare
172)	*Wallichia triandra*	Arunachal Pradesh	Endemic, rare due to lack of regeneration
Cyperaceae			
173)	*Carex kingiana*	Sikkim (Pheedong)	Endemic, known from type only
174)	*C. munroi*	H.P., *ca* 3450 m	Endemic, known from type only
175)	*C. praestans*	Kum., 2100-2400 m	Endemic, known from type only
Poaceae			
176)	*Agropyron duthei*	H.P., Kum., Garh.	Endemic, not collected since 1889
177)	*Calamogrostis garhwalensis*	J.K.,H.P.,Kum.,Garh., Nepal	
178)	*Deyeuxia kashmeriana*	J.K., Pakistan	
179)	*D. simlensis*	H.P.	Presumed extinct, known from Simla only
180)	*Elymus thomsonii*	H.P.,U.P.,Nepal	Not collected since 1883

	(1)	(2)	(3)
181)	*E. stewartii*	J.K.	Endemic
182)	*Festuca lucida*	J.K.,H.P.,Kum., Garh. upto 5000 m	Rare
183)	*Orinus thoroldii*	J.K. (Ladakh), Tibet upto 4700 m	
184)	*Poa falconeri*	H.P., Pakistan, Nepal, 3900 m	
185)	*P. setulosa*	H.P., Pakistan	
186)	*Pseudodanthonia himalaica*	H.P. (Kinnaur), 4000 m	Endemic
187)	*Puccinellia kashmiriana*	J.K. (Kameri Valley), H.P. (Lahul), 3500-5000 m	Rare, known only from few collections
188)	*Trisetum micans*	J.K., Garh.	Endemic

Jain, S.K. & A.R.K. Sastry (1980): *Threatened plants of India. A state-of-the-Art Report,* New Delhi.

Jain, S.K. & R.R. Rao (eds.) (1983): *Assessment of threatened plants of India.* BSI., Howrah.

Maheswari, J.K. (1980): In *Glimpses in Plant Research.* (P.K.K. Nair ed.) **5**:173-181. New Delhi.

Maheswari, J.K. (1984): In *The vegetational Wealth of the Himalayas.* (G.S. Paliwal ed.) Puja Publishers, Delhi.pp. 256-265.

Nayar, M.P. (1977): *Bull. Bot. Surv. India.* **19** (1-4):145-155.

Nayar, M.P. & A.R.K. Sastry (eds.) (1987-1990): *Red Data Book of Indian Plants* 3 Vols. BSI, Calcutta.

Polunin, O. & A. Stainton (1984): *Flowers of the Himalaya.* Oxford Univ. Press, New Delhi.

Raizada, M.B. (1977): *Bull. Bot. Surv. India.* **19**(1-4):174-175.

Raizada, M.B. (1980): *Indian J. For.* 3(**4**):370-371.

Rao, R.R. (1994): *Biodiversity in India (Floristic aspects).* Bishen Singh Mahendera Pal Singh, Dehra Dun.

Rau, M.A. (1981): In *The Himalayas-aspects of change* (J.S. Lall ed.) Oxford Univ. Press, New Delhi. pp. 50-63.

Rawat, J.K., Semwal, J.K. & A.N. Purohit (1985): *Blossoming Garhwal Himalaya.* HAPPRC, Garhwal Univ., Srinagar.

Sahni, K.C. (1973): *Proc. 12th Forestry Confr.,* Dehra Dun. 1-5. Appendix I-II.

Sahni, K.C. (1980). In *Glimpses in Plant Research* (P.K.K. Nair ed.) **5**:173-181, New Delhi.

Santapau, H. (1971). *IUCN Publication N.S. No.* **18**:115-126.

Shaffer, M.L. (1981). *Biosciences* **31**:131-134.

Active Phytomonitoring of Urban Ecosystem of Bombay

N.C. Joshi, S.M. Karmarkar and S.B.[1] Chaphekar

Department of Botany, R.J.College, Bombay, 400 088.

University of Pune and Poona 411 007.

INTRODUCTION

For more than two decades, we are conscious of the poor quality of our atmospheric environment, especially in industrial areas of cities. Bombay is acknowledged as one of the more polluted of these cities (Table 1). Different parts of the city face different levels of air pollution stress, which is reflected in health conditions of local residents. (Several reports by a leading Bombay NGO, SOCLEEN). Though periodic instrumental monitering of some important air polllutants done by agencies like the Greater Bombay Municipal Corporation (Table 2), it proves inadequate in two respects:- i) instruments cannot be operated all the time at all sites of concern., and ii) the value obtained fail to bring out their significance to life value since they do not account for accompanying phenomena like variations in ambient temperatures and relative humidity levels, wind directions and velocities or even interactions leading to synergistic or successive effects of different pollutants.

Plants, being stationary living organisms; are suitably equipped to play this role (Helmut, 1973). As many as seven different types of effects of air pollutants, on ecosystems have been categorized by Guderian (1977). It is recognized, however, that all plants do not respond equally or identically to all types of pollutants. Hence, it becomes necessary to study carefully the plant species selected as a candidate for phytomonitoring exercises. Consistency of response to a stress and clarity of expression of response are

Table 1: Quality of air at some areas of major cities of India

City (Sampling area)	Suspended particulate matter, mg/m^3 (24 hr/av)	SO_2, mg/m^3 (24 hr/av)	Sulphation rate, mg_3 $SO_3/100$ cm^3/day (Daily av)
Bombay (Lal Baug)	275	83	1.18
Calcutta (Cossipore)	413	48	0.60
Delhi (Najafgarh)	117	30	0.44
Hyderabad (Noulah)	236	27	0.66
Jaipur (Jhotwara)	226	11	0.12
Kanpur (Lajpatnagar)	261	20	0.18
Madras (Trivatrivur)	169	25	0.10

Source : NEERI, 1971.

Table 2: Quality of air over different parts of Greater Bombay city during the year 1986 in $\mu g/m^3$

Site	Pollutants	Jan.	Feb.	Mar.	Apr.	May	June	July	Aug.	Sept.	Oct.	Nov.	Dec.
Sion	SO_2	41	33	33	27	22	22	8	12	17	31	53	41
	NO_2	65	63	46	34	26	28	20	19	39	52	62	62
	SMP	363	352	363	279	245	196	206	155	185	278	324	336
Dadar	SO_2	50	34	36	39	36	60	19	11	11	34	31	53
	NO_2	67	68	42	43	21	26	20	25	30	48	53	70
	SMP	323	338	315	253	273	190	162	99	125	300	298	165
Goregaon	SO_2	12	9	16	19	13	8	7	6	7	17	24	13
	NO_2	30	22	14	12	11	8	7	9	17	26	31	37
	SMP	460	433	428	156	343	172	211	148	195	308	-	406
Ghatkopar	SO_2	75	35	46	44	32	13	6	26	37	33	N.A.	66
	NO_2	51	48	45	35	25	17	14	26	38	39	N.A.	55
	SMP	460	433	428	156	343	172	211	148	195	308	-	406
Colaba	SO_2	39	13	20	17	16	8	7	7	7	6	14	10
	NO_2	64	71	43	399	27	32	14	20	29	30	52	77
	SMP	264	210	224	154	205	126	151	102	115	158	284	269
CIBA*	SO_2	N.A.	39.4	N.A.	1.32	1.82	N.A.	N.A.	N.A.	N.A.	N.A.	N.A.	72.5
	NO_2	4.75	7.35	12.5	5.8	3.76	2.2	N.A.	N.A.	N.A.	N.A.	N.A.	7.15
	SMP	352	352	332	428	349	261	N.A.	N.A.	N.A.	N.A.	N.A.	316
Parel	SO_2	44	31	22	46	27	25	30	47	42	47	49	29
	NO_2	63	64	47	45	35	32	35	35	41	70	78	104
	SMP	463	507	413	318	322	146	183	152	216	354	411	426
Bombay Average	SO_2	37	24	27	29	21	19	11	18	17	24	28	29
	NO_2	53	49	38	32	24	26	21	20	31	41	45	55
	SMP	382	354	351	276	271	174	186	137	161	289	312	348

Source: i) Air quality monitoring cell, Khar, M.C.G.B.

* Hindustan CIBA GIEGY LTD, Readings of 1986-87.

N.A. = Not Available.

great significance in such cases. Posthumus (1983) has listed essential characteristic features of an indicator species. A plant species *Tithonia diversifolia* Hemsl. (Asteraceae) was used to assess the quality of environment of industrial-commercial city of Bombay that is also known for high population density.

MATERIALS AND METHODS

Tithonia diversifolia Hemsl. (Family Asteraceae)

Perennial herb, with striate stems, sparsely pubescent and glabrescent. Leaves alternate, unlobed, petiolate, 2-10 cm long, broadly winged almost to the base, biauriculate. Leaves pubescent with crenate-serrate margins. Fruit a dark brown achene.

Distribution

The plant, originally a native of tropical South America, has been introduced in India, Burma, Malaysia, China and South Africa. It is grown in the gardens of Bombay.

Seeds of *T. diversifolia* were obtained from a local dealer and sown in a garden plot. Seeds obtained from this generation were used for all the experiments.

Preparation of a Set

Plants were grown in polythene bags containing soil and farmyard manure in 3:1 ratio. Three plants were grown from seeds, in each bag and 5 such bags were prepared. This constituted one set. The plants in 3-4 leaf stage were transported to the sites selected, in different part of the city.

Description of Sites

Sets of plants were exposed to ambient air in residential areas, industrial areas and roadsides. A detailed description of the sites and the schedule of exposures are given in Table 3. Location of the sites can be seen on the map of Greater Bombay city (Plate 1). Plants were watered by volunteers during the 30 day exposure and above-ground phytomass of each plant was determined after harvesting and drying in oven at $60^{o} \pm 1^{o}C$.

Exposed plants were noted for any visible injury. Meteorlogical data, especially on temperature and relative humidity were collected from two stations of the India meteorology Department. Wind roses, based on 10 year averages, were also collected from IMD for two times a day. Tables 4 and 5 give temperature and R.H. conditions in north and south Bombay respectively during the period of studies. Wind roses are presented in plates 2 and 3.

Results

The highly industrialised and densely populated city of Bombay provides ideal conditions for the study of the effects of urban stresses on plants. The type of air pollutants, and their levels vary from place to place and season to season. This condition has arisen due to scattered industries, high vehicle density and pollution caused by population.

Plants are sensitive to air pollutants bur their sensitivities differ from one species to another (Farooq and Beg, 1980). The effects of air pollution are broadly divided into acute effects leading to severe foliar injury and chronic effects resulting due to long exposures at low concentration and culminating in a invisible injury such as reduction in growth, emaciation and ultimately total death of

PLATE 1 Map of Greater Bombay showing sites selected for the present study.
Sites 1 to 14 (Table 3)

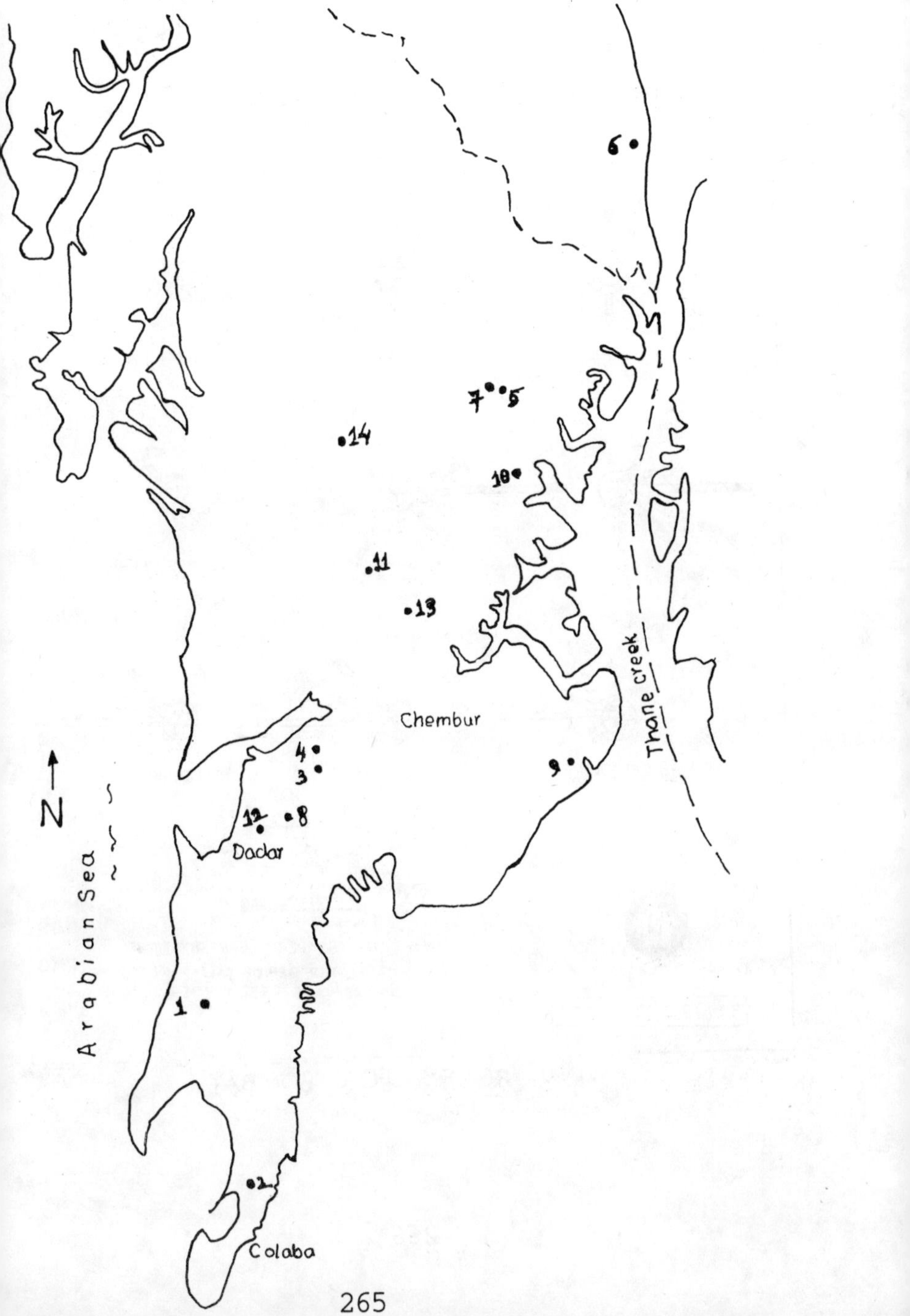

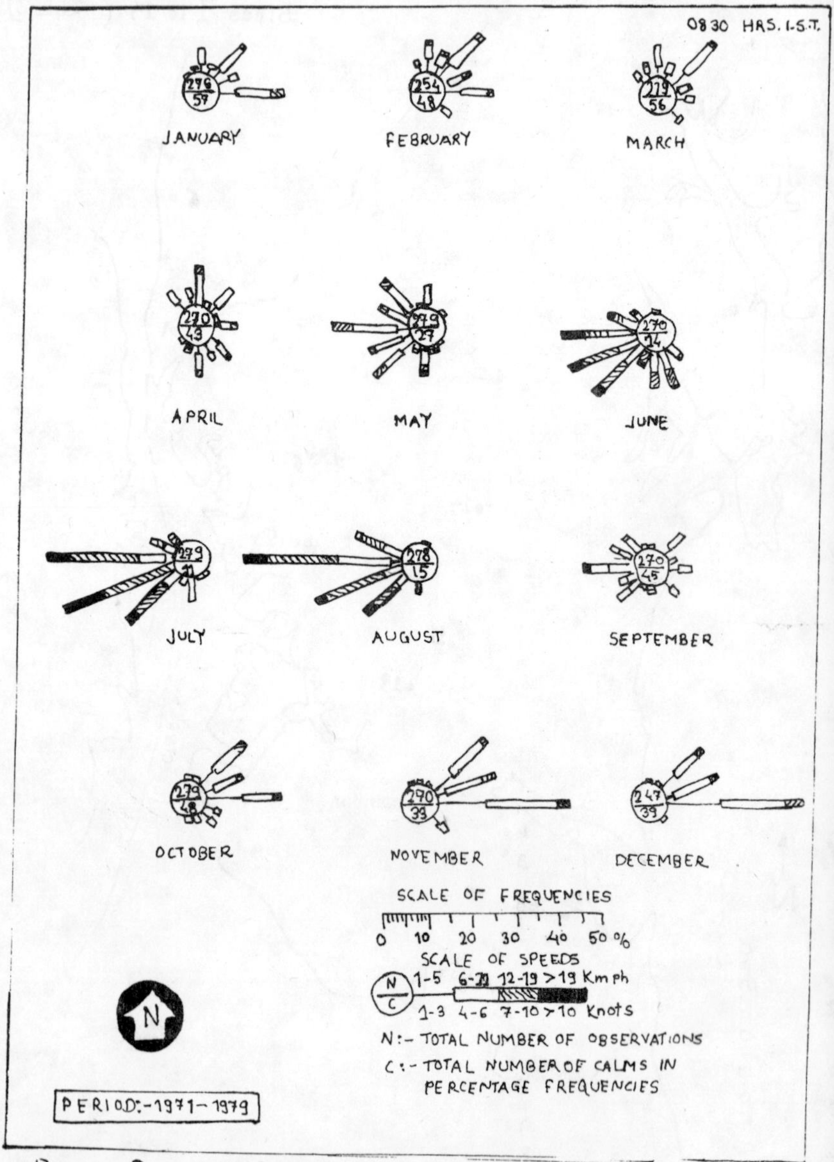

PLATE 2: WIND ROSES FOR BOMBAY

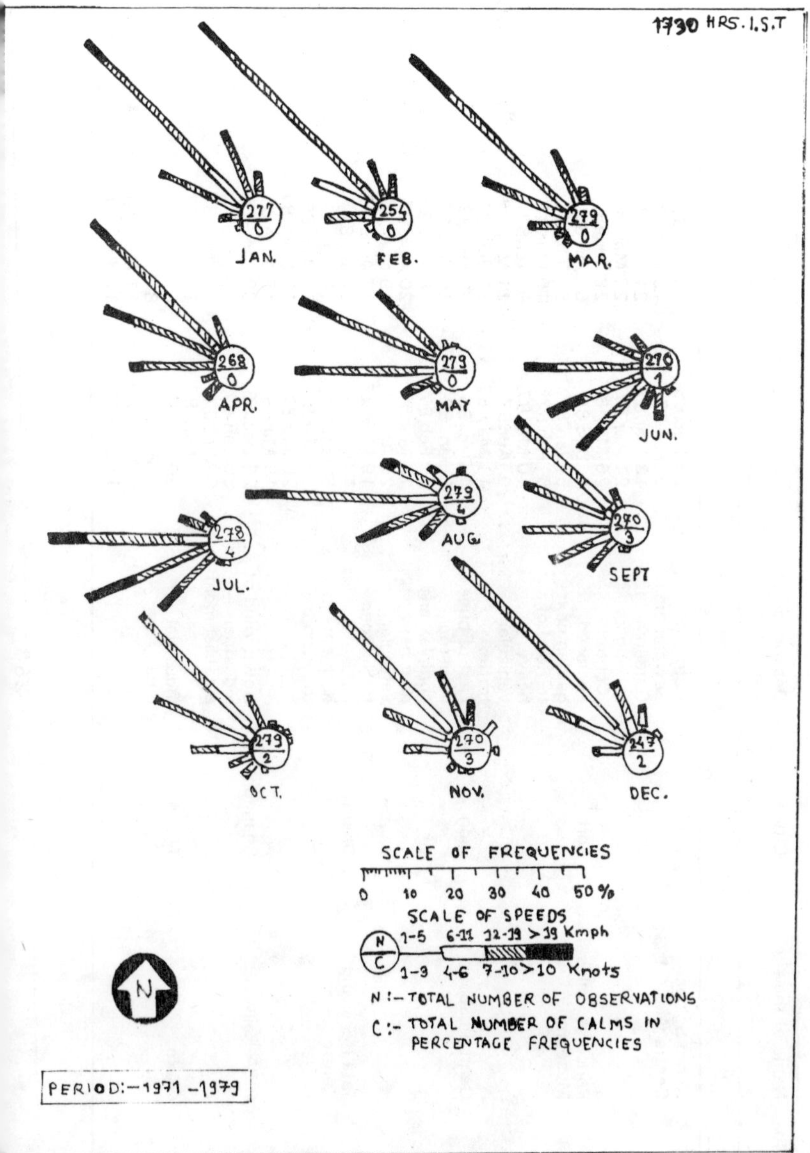

PLATE 3: WIND ROSES FOR BOMBAY

Table 3: Sites in Greater Bombay region chosen for phytomonitoring of air quality, using *Tithonia diversifolia*.

Sl.No.	Name of the site	City area	Description of the site	Schedule of exposure
1.	2.	3.	4.	5.
1.	Nehru Scinec Centre (N.S.C.)	Worli	Hutments, Sewage drain, Garden area serving as 'Control'	29 Dec.86 - 29 Jan.87 6 Feb.87 - 13 Mar.87 13 Mar.87 - 21 Apr.87 21 Apr.87 - 26 May 87 26 May 87 - 26 June87
2.	**Prince of Wales Museum**	Fort	Enclosed garden next to Road of heavy vehicular traffic	4 Dec.86 - 5 Jan.87 5 Jan.87 - 5 Feb.87 5 Feb.87 - 10 Mar.87 10 Mar.87 - 9 Apr.87 9 Apr.87 - 11 May 87 11 May 87 - 11 June87
3.	Central Labour	Sion	East of huge industrial complex and near heavy vehicular traffic road	4 Dec. 86 - 8 Jan. 87 8 Jan. 87 - 10 Feb. 87 10 Feb.87 - 18 Mar. 87 18 Mar.87 - 20 Apr. 87 20 Apr.87 - 19 May 87 19 May 87 - 17 June 87
4.	Tata Receiving Centre	Dharavi	Industrial, hutments in the vicinity	4 Dec. 86 - 8 Jan. 87 8 Jan. 87 - 10 Feb. 87 18 Mar.87 - 20 Apr. 87 20 Apr.87 - 15 May 87 15 May 87 - 17 June 87
5.	CIBA-Railway	Bhandup	Down-wind of industrial establishment	26 Dec.86 - 30 Jan. 87 30 Jan.87 - 6 Mar. 87 6 Mar. 87 - 11 May 87 11 May 87 - 17 June 87
6.	Glaxo	Thane	Industrial area	30 Jan.87 - 6 Mar. 87 6 Mar. 87 - 8 Apr. 87 8 Apr. 87 - 11 May 87 11 May 87 - 17 June 87

1.	2.	3.	4.	5.
7.	CIBA-Roadside CIBA (RS)	Bhandup	Upwind of Pharmaceutical industry,next to heavy vehicular traffic	6 Mar.87 - 8 Apr. 87 8 Apr.87 - 11 May 87 11 May 87 - 17 May 87
8.	U.D.C.T. (University Department of Chemical Technology)	Matunga	Residential area emissi-ons from the chemical department as well	13 Apr. 87 - 19 May 87 19 May 87 - 23 June 87
9.	Tata Electric (Tata TR)	Mahul	Industrial area inclu-ding huge petroche-mical industries	13 Apr. 87 - 19 May 87 19 May 87 - 26 June 87
10.	Godrej Colony	Vikhroli	Residential colony of an industrial complex, mainly pharmaceutical	13 Apr. 87 - 19 May 87 19 May 87 - 23 June 87
11.	Kalina (University campus)	Kalina	Nearest Road about 300 M away polluting industries around	19 Apr. 87 - 29 May 87 29 May 87 - 3 July 87
12.	Ruparel College	Matunga	College Play-ground. Next to industries	29 Apr.87 - 29 May 87 29 May 87 - 3 July 87
13.	Somaiya College	Vidyavihar	College campus Resi-dential area	24 Apr. 87 - 26 May 87 26 May 87 - 26 July 87
14.	Municipal Garden, Jogeswari	Jogeswari	Alongside vehicular traffic and highway	29 Apr.87 - 29 May 87 29 May 87 - 3 July 87

Table 4: Temperature and relative humidity recorded at the Indian Meterological Centre, Santacruz, Bombay

Month	Temperature		Relative humidity		
	Mean Max.	Mean Min.	Mean Max.	Mean Min.	Mean
Dec. 86	31.6	17.3	79.0	30.0	54.5
Jan. 87	31.3	18.1	85.7	35.0	60.3
Feb. 87	32.2	19.0	79.0	27.3	51.0
Mar. 87	33.4	21.9	85.2	35.2	60.2
Apr. 87	33.8	24.3	84.8	44.5	64.5
May 87	34.2	26.2	83.1	48.06	65.5
June 87*	32.6	25.8	89.09	66.7	77.8

* Total number of rainy days in June = 14

Table 5: Temperature and relative humidity recorded at the Indian Meterological Centre, Santacruz, Bombay.

Month	Temperature		Relative humidity		
	Mean Max.	Mean Min.	Mean Max.	Mean Min.	Mean
Dec. 86	31.4	19.6	80.0	42.0	61.5
Jan. 87	30.7	20.2	84.2	48.0	66.1
Feb. 87	31.4	21.8	87.0	43.0	65.0
Mar. 87	32.4	23.3	83.3	52.2	67.7
Apr. 87	33.3	25.2	80.06	54.3	67.0
May 87	34.1	27.0	79.7	53.8	66.0
June 87*	32.4	26.1	89.1	70.0	79.0

* Total number of rainy days in June = 14

the plant organs (Posthumus,1983). Exposure of plants to pollutants at high concentration for long periods, generally results in the development of symptoms of visible injury and associated physiological disturbances. These responses are largely irreversible and may lead to reduction in plant growth development and yield. In contrast, visible injury does not result when exposures are short or pollutant concentrations are low, but physiological disturbances and growth reductions are frequently observed (Black, 1984).

In the general scheme of plant responses to air pollution, it is recognised that many factors, both pollutants and environmental, influence relative sensitivity of the plants. The factors influencing the plant response to air pollution include the toxicity of pollutant, concentration, duration of exposure, stomatal behaviour, pollutant uptake by the plant and prevailing environmental conditions. In addition, response of plants are also modified by the age of the tissue examined, nutrient status and the species or variety of the plant (Black, 1984).

The experimental plants were examined and the following observations were made:

Acute damage to 2-3 individual plants leading to their death was observed at Sion (C.L.I.), in Jan.-Feb.-Mar. indicating acute pollution doses suffered by plants at that site. A number of individuals showed margin necrosis of young leaves in Jan.-Feb., at Museum and Dharavi sites. Same period was also bad for plants at Sion (C.L.I.) and Bhandup (CIBA-RT) as shown by chlorotic patches and tip-margin burns respectively. Feb.-Mar. period also was injurious to plants at Sion, Thane and Bhandup. April-May period was not good at Thane.

Apart from visible injury, other type of injuries, *i.e.* invisible ones, also could be perceived.

Injury to plants due to air pollution is likely to be of three types.

i) Death and elimination of the plant due to servere injury affecting all parts of the plant. Plant biomass in such cases is recorded as nil.

ii) Injury to some parts, especially to leaves of plants, exhibiting characteristic symptoms. Most of the plants except for their injured leaves were known to recover from this type of injury,provided the pollutant was withdrawn. In such cases plant productivity is reduced in proportion to the injury suffered (Boralkar, 1979) or it may be altered depending upon the stage of the growth (Van Haut, 1968).

iii) It is possible that a plant subjected to pollution may not show any visible injury due to pollution. An invisible injury may be perceived in the form of reduced flower, fruit and seed formation and decrease in productivity and yield (Manning and Fedder, 1980). There are several versions of the effects on biosynthesis in pollution-affected plants that lead to reduction in yields (Chaphekar and Boralkar, 1979; Pell and Pearson, 1984). Some of these steps are not fully understood so far. It is generally agreed, however, that adverse effects of pollutants in plants are perceptible in the form of a decline in net primary productivity and protein synthesis and sometimes an increase in the sugar content of the plant (Agarwal, 1985).

Phytomass of *T. diversifolia*

During the period of January-February, plants grown at all the sites showed a significantly low above ground phytomass with the lowest being recorded at sites in Dharavi [Tata (D)], followed by Sion (C.L.I.). In the next period (February-March) also, plants grown at all the sites showed a significantly low phytomass value as compared to that at the control site *i.e.* Worli (N.S.C.). The lowest phytomass during the above period was recorded at Sion (C.L.I.) followed by that in plants grown at Thane (Glaxo). During the period of March-April the plants grown at the sites showed a significantly reduced phytomass in comparison to the values of phytomass in plants at Worli (N.S.C.). The lowest phytomass was observed in the plants from Bhandup (CIBA-RT) followed by that at Thane (Glaxo) (Table 6).

In the period of April-May plants grown at seven sites showed a significant difference in the above-ground phytomass from the phytomass obtained from site, Worli (N.S.C.). The plants, at all the above seven sites, showed significantly reduced phytomass. During the above period lowest phytomass was observed at Museum followed by those at Trombay (Tat-Tr). During the months of May-June plants grown at ten sites showed significantly low phtomass as compared to the control. A significant reduction in the phytomass was noted at sites Bhandup (CIBA-RT), Museum, Dharavi [Tata (D)], Thane (Glaxo), etc.

During the entire study period, a significantly low phytomass was observed in plants grown at thirty one different sites over different periods which was 80% of the total values taken during the study period.This suggests that the above ground phytomass of *T. diversifolia* can be regarded as good indicator of air pollution.

Table 6: Above-ground phytomass (g. dry weight plant^{-1}, of *Tithonia diversifolia*, at different sites in Bombay city. Plants were exposed to ambient air from December 1986 to June 1987, (Mean of 12 samples ± S.E.)

Site No. and Name	Dec.-Jan.	Jan.-Feb.	Feb.-Mar.	Mar.-Apr.	Apr.-May	May-June
1. N.S.C.	N.P.	6.066 ± 0.283	9.965 ± 1.248	8.286 ± 0.288	4.015 ± 0.617	6.021 ± 0.86
2. Museum	2.603 ± 0.275	*3.375 ± 0.592	*3.736 ± 0.416	*2.940 ± 0.200	*1.85 ± 0.335	*2.040 ± 0.630
3. C.L.I.	2.574 ± 0.538	*1.843 ± 0.216	*1.075 ± 0.141	N.P.	3.872 ± 0.285	*4.154 ± 0.288
4. Tata (D)	3.765 ± 0.465	*1.613 ± 0.276	*5.715 ± 0.723	*3.162 ± 0.604	4.464 ± 0.677	*2.098 ± 0.570
5. CIBA (RT)	N.P.	*2.102 ± 0.230	*4.313 ± 0.449	*1.672 ± 0.164	*2.291 ± 0.350	*1.894 ± 0.288
6. Glaxo	N.P.	N.P.	*2.637 ± 0.282	*3.549 ± 0.483	*2.250 ± 0.286	*2.141 ± 0.399
7. CIBA(RS)	N.P.	N.P.	N.P.	*2.472 ± 0.514	*2.545 ± 0.166	*3.281 ± 0.308
8. U.D.C.T.	N.P.	N.F.	N.P.	N.P.	*2.347 ± 0.237	4.120 ± 0.877
9. Tata (Tr)	N.P.	N.P.	N.P.	N.P.	*2.197 ± 0.325	5.164 ± 1.040
10. Godrej	N.P.	N.P.	N.P.	N.P.	5.390 ± 0.645	*2.239 ± 0.282
11. Kalina	N.P.	N.P.	N.P.	N.P.	*2.481 ± 0.393	Sample inadeqate
12. Ruparel	N.P.	N.P.	N.P.	N.P.	4.293 ± 0.586	*3.893 ± 0.398
13. Somaiya	N.P.	N.P.	N.P.	N.P.	3.664 ± 0.477	*2.144 ± 0.295
14. Jogeshwari	N.P.	N.P.	N.P.	N.P.	4.313 ± 0.840	*3.136 ± 0.147

N.P. = Not placed. * = Significant at $P > 0.05$

Table 7: Percentage variations in above-ground phytomass of *T. diversifolia* due to its exposure to ambient air at different sites in Bombay city during January 1987 to June 1987.

Site No. and Name	Jan.-Feb.	Feb.-Mar.	Mar.-Apr.	Apr.-May	May-June
1. N.S.C.	0	0	0	0	0
2. Museum	-44.36	-62.50	-64.51	-53.92	-66.11
3. C.L.I.	-69.61	-89.21	N.P.	-3.56	-31.00
4. Tata (D)	-73.40	-42.64	-61.83	+11.18	-65.15
5. CIBA (RT)	-65.34	-59.71	-79.82	-42.93	-68.54
6. Glaxo	N.P.	-73.53	-57.16	-43.96	-64.44
7. CIBA (RS)	N.P.	N.P.	-70.16	-36.61	-45.50
8. U.D.C.T.	N.P.	N.P.	N.P.	-41.54	-31.57
9. Tata (Tr)	N.P.	N.P.	N.P.	-45.28	-14.23
10. Godrej	N.P.	N.P.	N.P.	+34.24	-62.81
11. Kalina	N.P.	N.P.	N.P.	-38.20	Sample inadequate
12. Ruparel	N.P.	N.P.	N.P.	+6.92	-35.34
13. Somaiya	N.P.	N.P.	N.P.	-8.74	-64.39
14. Jogeshwari	N.P.	N.P.	N.P.	+7.42	-47.91

N.S.C. = Control

Table 8: Air quality as depicted by above-ground phytomass of *T. diversifolia*

Site No. and Name	Jan.-Feb.	Feb.-Mar.	Mar.-Apr.	Apr.-May	May-June
1. N.S.C.	C.	C.	C.	C.	C.
2. Museum	M.A.	V.A.	V.A.	M.A.	V.A.
3. C.L.I.	V.A.	V.A.	N.P.	P.C.	A.
4. Tata (D)	V.A.	M.A.	V.A.	C.	V.A.
5. CIBA (RT)	V.A.	M.A.	V.A.	M.A.	V.A.
6. Glaxo	N.P.	V.A.	M.A.	M.A.	V.A.
7. CIBA (RS)	N.P.	N.P.	V.A.	A.	M.A.
8. U.D.C.T.	N.P.	N.P.	N.P.	M.A.	A.
9. Tata (Tr)	N.P.	N.P.	N.P.	M.A.	P.C.
10. Godrej	N.P.	N.P.	N.P.	C.	V.A.
11. Kalina	N.P.	N.P.	N.P.	A.	Sample inadequate
12. Ruparel	N.P.	N.P.	N.P.	C.	A.
13. Somaiya	N.P.	N.P.	N.P.	P.C	V.A.
14. Jogeshwari	N.P.	N.P.	N.P.	C.	M.A.

i) 0% and more towards positive side = Clean (C)
ii) 0 to -20% = Partially clean (P.C.)
iii) -20 to -40% = Acute (A)
iv) -40 to -60% = Moderately Acute (M.A.)
v) -60% and more towards side = Very Acute (V.A.)

The degree of the health of plants is an index of the existing environmental conditions in urban areas and this can be judged from the above ground parts of the plant (Bhavanishankar, 1986). Reduced plant growth and yields are observed even in the absence of visible injury (Bell and Clough, 1973; Banerjee, 1976). This was also observed in *T. diversifolia* at several sites. When plants are exposed to subacute doses of air pollution, they tend to indicate the same by reduction in growth (Banerjee and Chaphekar, 1978; Rao, 1981).

Percentage variations of the values obtained in comparison with the corresponding values for plants grown at the control site, *i.e.* Worli (N.S.C.), were calculated for each duration of exposure and these are presented in Table 7. The status of air quality as revealed through these finding are as follows:

i) **January-February:** The most polluted site during this period is site Dharavi [Tata (D)] followed by Sion (C.L.I.).

ii) **February-March:** The most polluted site during this period is Sion (C.L.I.), followed by Thane (Glaxo).

iii) **March-April:** The most affected site during this period is Bhandup (CIBA-RT and RS).

iv) **April-May:** The most polluted site during this period is Museum followed by Trombay (TATA-TR), Thane (Glaxo), Bhandup (CIBA-RT) and King Circle (U.D.C.T.). The site Vikhroli (Godrej) shows good air quality, the phytomass value being slightly higher (though not significant) than that of the control site.

v) **May-June:** Poor air quality is observed at Bhandup (CIBA-RT), followed by that at Museum, Dharavi [TATA (D)] and Vikhroli (Godrej). The site Trombay (TATA-TR) shows good air quality, the phytomass value being slightly higher than that of the control site.

In an earlier work, growth as indicated by phytomass of *Helianthus annus* var. *Japonica* provided a reliable phytomonitor on a semiquantitative scale (Chaphekar *et al.*, 1980). The present work shows that *T. diversifolia* belonging to the same family is a good candidate for similar studies . This plant species can be used on larger scale for monitoring ambient air quality by putting differences in phytomass values obtained in Table 6, at different sites, on five point scale as suggested below:

i)	0% and more towards positive side	=	Clean(C)
ii)	0% to -20%	=	Partially Clean (P.C.)
iii)	-21% to -40%	=	Acute(A)
iv)	-41% to -60%	=	Moderately Acute (M.A.)
v)	-61% to above	=	Very Acute (V.A.)

The ambient air quality thus obtained is presented in Table 8. Some salient features of air quality in this city are as follows:

1. The sites Bhandup (CIBA-RT), Thane (Glaxo), and Museum show consistent pollution.

2. More than 50% of the areas tested in May-June show very acute pollution.

3. In the months of January-February to March-April all the sites showed pollution levels ranging between moderately acute to acute.

4. When the work was extended over more number of sites in April-May and May-June, even areas like Vidyavihar (Somaiya)and Kalina showed that they suffered from pollution.

5. Sites immediately next to vehicular sources of pollution *i.e.* Museum, Jogeshwari and Bhandup (CIBA-RS) suffer from very acute to moderately acute pollution in May-June.

It is pertinent to mention here that it is generally recognised that the entire city of Bombay suffers from air pollution at one level or another and that no site is really unpolluted. Choice of any one site as a control within the city, is hence, open to question. This peculiar situation of variability of pollution generation and dispersal conditions may therefore be responsible for observing some samples being more healthy than the controls at certain durations of exposure of plants. All the same, phytomonitoring proves to be a rewarding excercise as long as one does not overlook that the results of these excerises are to be taken as providing information about ambient air quality on a relative scale rather than absolute. The present excercise is an important step in standarizing phytomonitoring and is recommended as a reliable monitor for implementation during air pollution studies in this area on a large scale and for sustained periods in future.

AKNOWLEDGEMENTS

The authors are grateful to the authorities of the Institute of Science, Bombay, from where much of this work was carried out. Ministry of Environment and Forests, New Delhi, sponsored the work, which is gratefully acknowledged.

REFERENCES

Agrawal, M. (1985): In *Symp. Biomonitoring of States of Environmemt: Bioindicators:* for I.C.S.U. by R.L. Press. 1986.

Banerjee, A.K. and S.B. Chaphekar (1976): *Geobios* **5**:8-11.

Banerjee, A.K. (1976): *Plants and pollution.* M.Sc. thesis to University of Bombay.

Bell, J.N.B. and W.S. Clough (1973): *Nature* , London, **241**:47-49.

Bhavanishankar, V. (1986): *Studies of Plant Responses to Environmental Stress of Industrial Origin*, Ph.D. thesis, University of Bombay, Bombay.

Black, V.J. (1984): In *Gaseous Air Pollutants and Plant Metabolism.*: (Koziol, M.J.and Whatley, F.R. ed.) Butterworths, 223-248.

Boralkar, D.B. (1979): *A Contribution to the Study of the Effects of Air pollutants on Plants.* Ph.D thesis, University of Bombay, Bombay.

Chaphekar, S.B., Boralkar, D.B. and R.P. Shetye (1980): In: *Proc. V. Internal. Symp. Trop, Ecol. and Development,* (J.I. Furtado, ed.) Kualalumpur, 669-675.

Chaphekar, S.B. and Boralkar, D.B. (1976): *Ind. J.A.P.C.* **2** (1):19-23.

Farooq, Masood, A. and Beg, M.H. (1985): *Environ. Pollut.* (Seri A), **39**:197-205.

Guderian, R. (1977): *Air pollution phytotoxicity of acidic gases and its significance in air pollution control.* Springer-Verlag, New York.

Helmut, B. (1973): *Toxicology*, **1**(1):79-89.

Manning, W.J. and J. Fedder (1980): In *Biomonitoring Air pollution with Plants.* Appl. Sci. Publ. Ltd., London.

Pell, E.J. and N.S. Pearson (1984): *Environ. Pollut.* (Ser. A) **35**:345-352.

Posthumus, A.C. (1983): *Environmental Monitoring and Assessment*, **3**:263-274.

Rao, D.N. (1981). In *Proc. Workshop Biol. indicators of Environmental Pollution.* Osmania University, Hyderabad, 1-8.

Water Relations of Halophytes in Indian Arid Zone

D.N. Sen and S. Mohammed
*Laboratory of Plant Ecology,
Department of Botany,
J.N.V. University, P.O. Box. 14,
Jodhpur-342 001, India.*

INTRODUCTION

The most important phase of plant, water relations with respect to its effect on growth and development is the internal water balance. Water relations and stomatal behaviour are important indices which reflect the ability of plants to economise essential requirements under prevailing climatic and edaphic conditions (Jarvis, 1967; Sen *et al.*, 1972). The intensity of any factor operating in specific habitat has a great influence on the adaptive strategies in plant and the physiognomy of the vegetation. Saline ecosystem faces a wide range of environmental fluctuations, so plants have to develop more adaptations. The Indian arid zone is characterized by erratic and scanty rainfall, clear skies, high day temperature in summer, high evapotranspirations, saline and sandy soils, and poor vegetations. Intense solar radiations associated with high temperature, dryness of the air and shortage of soil moisture are most important climatic factors to which plants must adapt (Sen, 1982).

Water is one of the most important constituent of the multifactorial environment which exerts its operative influence on any and every ontogenic phase in the life cycle of plant. The survival of plant depends largely on its capacity for adaptive transformations of the course and the rate of physiological season (Oppenheimer, 1951). When the plants have more water in their tissues, then they have greater ability for protoplasm to withstand permanent injury during drought. As tissue becomes dessicated, protoplasm becomes increasingly dense and its viscosity gradually increases; when dehydration is severe, complete gelation of the protoplasm occurs and finally it may become rigid to the point of brittleness (Levitt, 1956). In saline environment, soil has multiple effects on plants and among them prominent one is the osmotic effect of high salt concentration. High amounts of soluble salts in soil solution form a considerable component of the total soil moisture stress. A critical aspect in determining the success of halophytic species in saline environments is their ability to lower water potentials, sufficiently to maintain normal water uptake relationships (Waisel, 1972; Sen & Mohammed, 1994). The main difficulty with which desert plants are confronted is to maintain a favorable balance between absorption and transpiration under the adverse conditions of environment. It has been stated that relationship between the relative water content (relative turgidity)and water potential of leaves measure the amount of water held by leaf tissues and varies considerably according to type, age of the plant and growing conditions (Stayer, 1960).

The plants of Indian arid zone are totally dependent upon the availability of water in the rainy season , and the water deficiency is the most important factor controlling their growth and survival. The rainy season water is depleted by the beginning of winter. Besides the lack of soil moisture, salt tolerance is an important property for all plants which, at any time during their life cycle, are exposed to the effect of salt. Such effects may be temporary, e.g. when roots pass through saline soil layers, or permanent as in the case of halophytes. The environment of arid saline ecosystem offers an adaptive challenge to the survival of plants, where only those species which possess mechanism to adjust themselves under the strong climatic fluctuations can survive, whereas others cannot. Since water is the dominant climatic factor, ecophysiology of water relations become one of the most important parameters to judge the capability of a particular species to adjust itself under prevailing climatic and edaphic conditions. Therefore, comparative studies were done here on osmotic potential and water relations of halophytes at saline site-I (Pachpadra; Electric conductivity ranges from 3.2-13.0 mmhos/cm. at 1:5 soil-water extract) and non-saline site-II (Jodhpur; EC:0.18-2.1 mmhos/cm.) to determine the monthly variations in the state of water balance and its relationships with soil moisture conditions.

MATERIALS AND METHODS

The present study was carried out throughout the year during 1986-1988, depending on the availability of plants at sites-I&II. For the present investigation, saline plants such as *Salsola baryosma* (Roem, et Schult.) Dandy, *Sesuvium sesuvioides*(Fenzl.) Verdc., *Suaeda fruticosa* (Linn.) Forsk. and *Trianthema triquetra* Rotter ex Willd. were found at both the sites, while *Cressa cretica* Linn. and *Zygophyllum simplex* Linn. are found at site-I, were selected. With the termination of the monsoon season, the plants of *S. sesuvioides* disappeared after completing it's life cycle in the months of November and December at sites I & II, respectively. Likewise the plants of *T. triquetra* and *Z. simplex* at site I were not observed, as they also complete their life cycle by the end of December and February, respectively. The observations were recorded at the middle of each month. For the measurement of fresh weight, plant sample was immediately weighed and then kept in an oven at 80° for 24 h for assessing its dry weight. Osmotic Potential (OP; - bars) of leaves was determined as per Janardhan *et al.* (1975), while Relative Water Content (RWC) by Weatherley (1950), and Bound water (BW) by Schonbeck & Norton (1979). Leaf Water Content (LWC) of leaves was determined on the dry wt. basis as follows:

$$LWC = \frac{\text{Fresh wt. - Dry wt.}}{\text{Dry wt.}} \times 100$$

For soil moisture, soil samples were collected monthly from the two depths: surface and 20 cm at both sites and calculated following the method given by Piper (1950). The data are presented here seasonally. The obtained data were statistically analysed after Gomes & Gomes (1984).

Observations

(a) **Osmotic Potential (OP)**

It is evident from Fig.1 that the values of osmotic potential of leaves of different saline plants revealed that the lowest value of OP (-20 bars) was obtained from *S. sesuviodes* at site-I in July, while -34.5 in *T. triquetra* at site-II in January. In *S. baryosma* the OP was highest at both sites, *i.e.* -107.5f and -110.0 bars at sites -I & II, respectively, in the month of December, followed by *S. fruticosa* at sites-I (90.0 bars) & II (-77.5 bars) in the month of April and December, respectively. These higher values in leaves were due to continuous uptake of soluble ions by plants. *C. cretica* and *Z. simplex* exhibited higher values of OP, *i.e.* -85.0 & -83.5 bars during April & December, respectively at site-I. The data were significant at 5% probability level for all species at both sites.

(b) **Leaf Water Content (LWC)**

It reveals from Fig.2 that leaf water content was high in July-August in all the species at both sites. Lowest LWC was observed during dry period. The leaves of succulent halophytes, such as *Z. simplex* (1809%), *S. sesuviodes* (1315% at site-I and 1213% at site-II), *S. triquetra* (990% at site-I and 1250% at site-II) and *S.*

MONTHLY VARIATIONS IN OP (-bars) OF HALOPHYTES AT SITES - I & II.

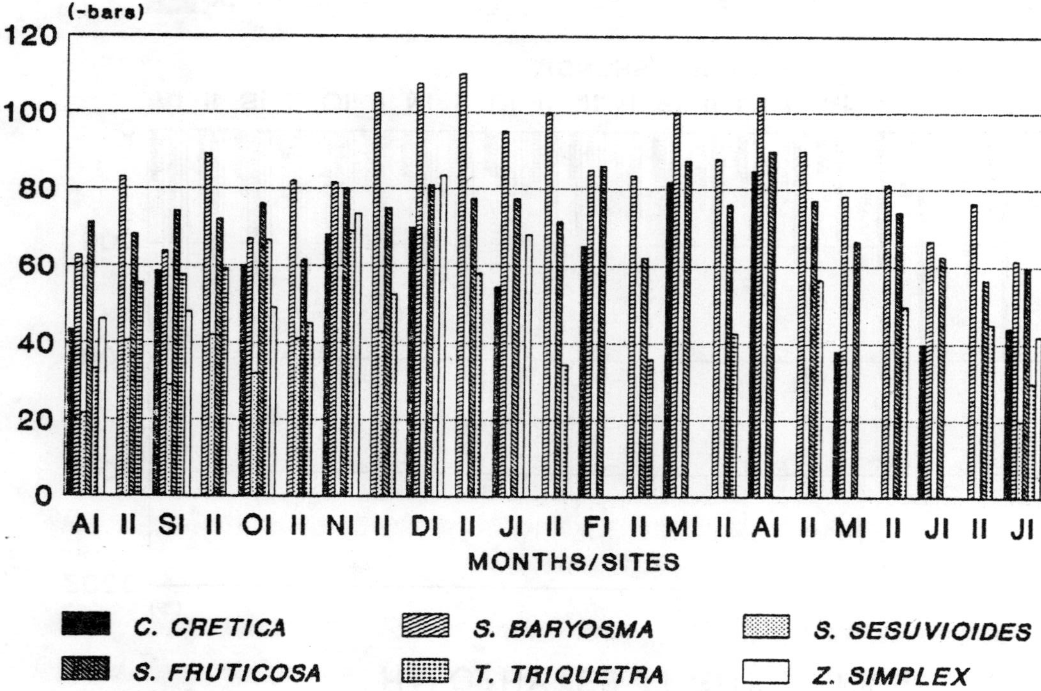

MONTHLY VARIATIONS IN LWC (%) OF HALOPHYTES AT SITES - I & II.

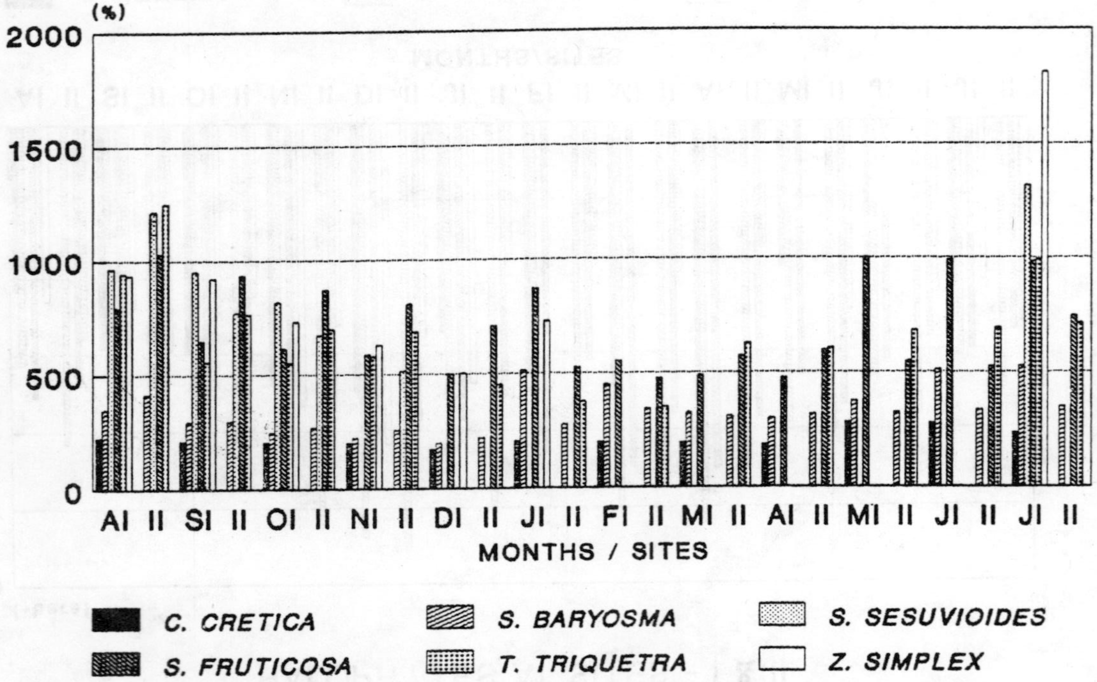

baryosma (524% at site-I and 415% at site-II) contain more LWC as compared to non-succulents like, *C. cretica* (286.5% at site-I). Almost all species, which grow at site-II exhibited more water content than from site-I. *S. baryosma, C. cretica* and *Z. simplex* showed lower values during December at site-I while, *S. fruticosa* and *T. triquetra* during February at site-II. The data were significant at 5% level for all species.

(c) **Relative Water Content (RWC)**

Fig. 3 shows that relative water content was highest during rainy months and lowest in dry in all species at both sites. Among all plant species, *S. fruticosa* showed highest (104.7% at site-I and 99.9% at site-II) RWC at site-I. However, *S. baryosma, S. sesuvioides* and *T. triquetra* from site-II exhibited more RWC than site-I. *S. fruticosa, S. baryosma, S. sesuvioides, T. triquetra* and, *Z. simplex* showed highest RWC during July, while *C. cretica* during January at site-I. The data were significant for all species at both sites.

(d) **Bound Water (BW)**

Bound water in leaves of plant species at two different sites are given in Fig.4. The results reveal that *S. fruticosa* at site -I, exhibited maximum (95%) value in April and minimum (33%) was in September. The minimum value among studied plants was in *C. cretica.* Generally, all plant species showed higher values during saline and water stress period and lower during moist period. The data were significant for all species at both sites.

(e) **Soil Moisture (SM)**

The soil moisture study showed that the soil moisture was maximum during rainy season at both sites and later decreased (Fig. 5). It was lower at the surface and was maximum at 20cm depth.

MONTHLY VARIATIONS IN RWC (%) OF HALOPHYTES AT SITES - I & II.

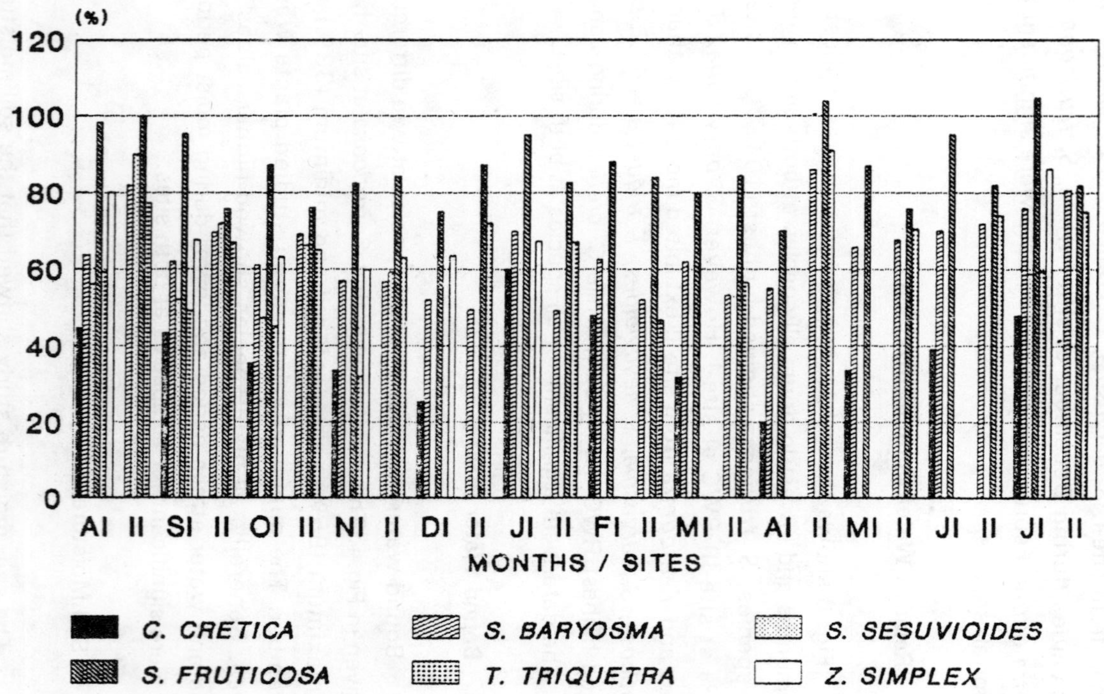

FIG. 3.

MONTHLY VARIATIONS IN BW (%) OF HALOPHYTES AT SITES - I & II.

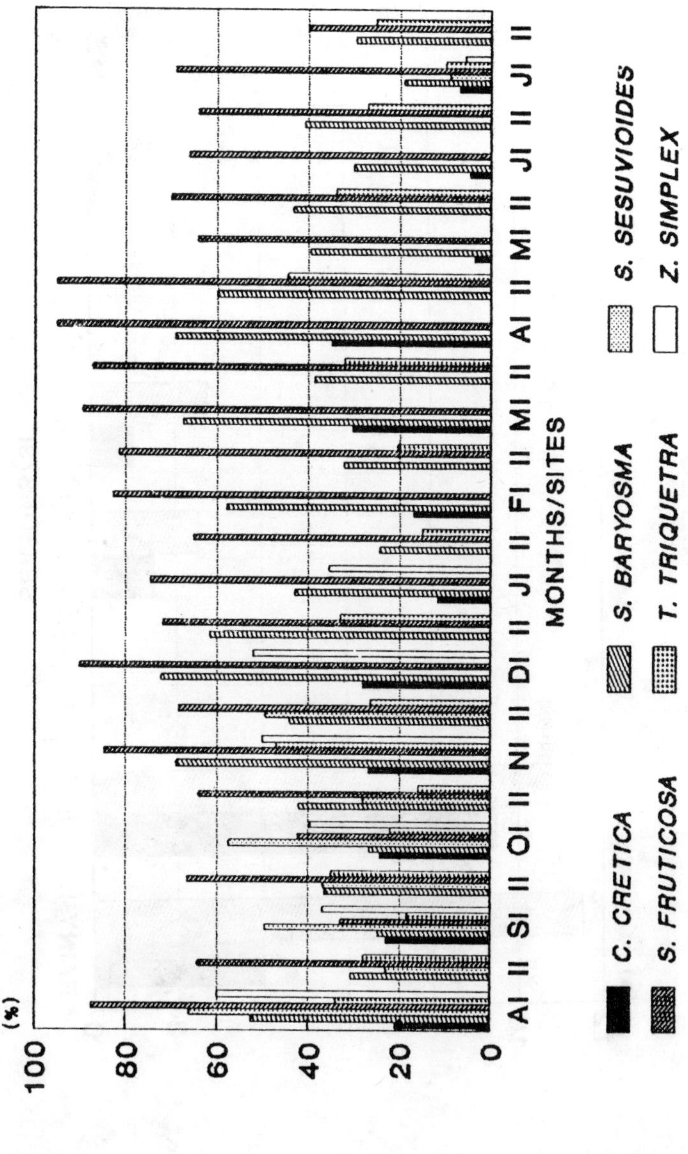

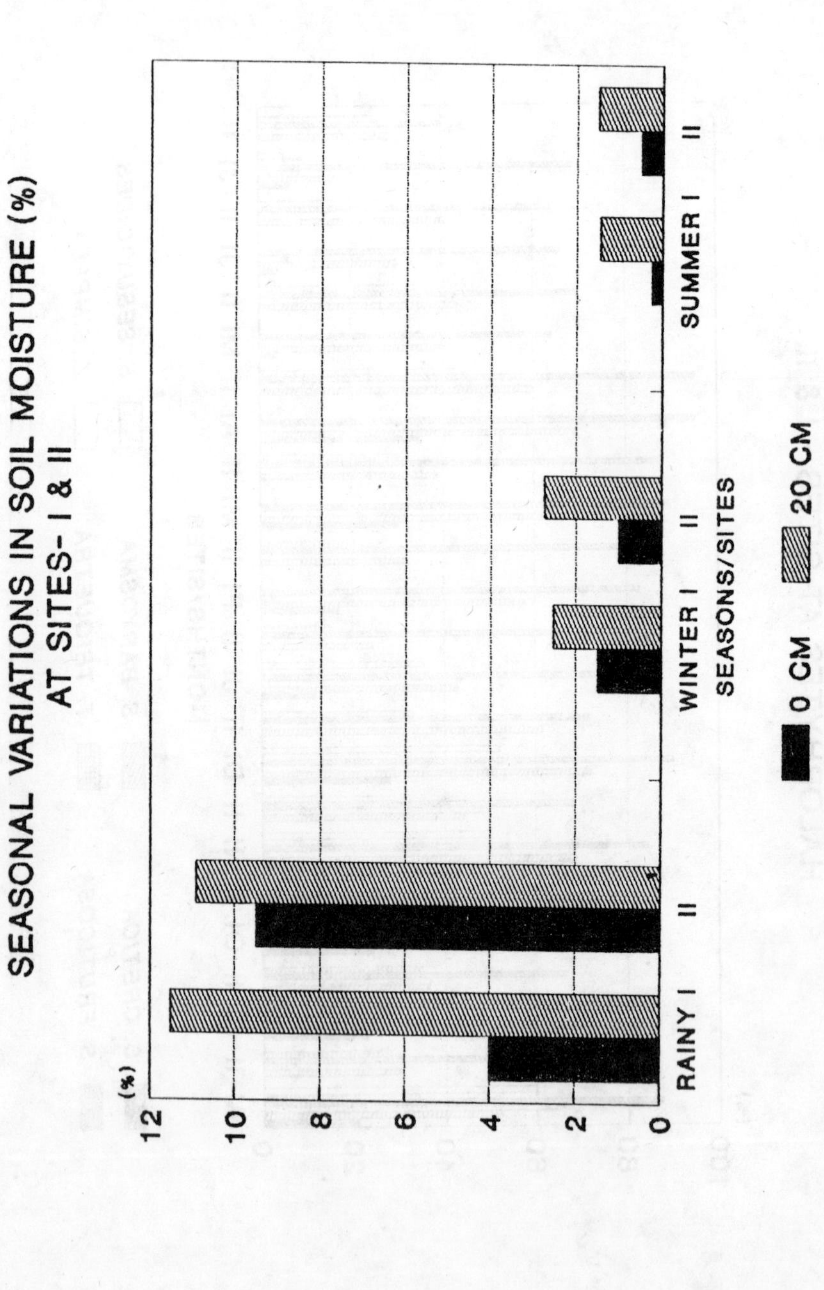
SEASONAL VARIATIONS IN SOIL MOISTURE (%)
AT SITES- I & II
(%)
12
10
8
6
4
2
0
RAINY I
II
WINTER I
II
SUMMER I
II
SEASONS/SITES
0 CM
20 CM

FIG. 5.

The trend of moisture content showed that the values decreased from rainy to summer season. The data were significant for both depths at both sites.

DISCUSSION

Ecophysiology of water relations is one of the important parameters to judge the capabilities of a particular species to adjust itself under prevailing climatic and edaphic conditions. Water balance of a species can be measured through its effect upon water content or water potential of the plant. A negative water balance always eventually produces a decrease in turgidity and water potential of the tissues (Larcher, 1983). These changes first appear in the leaves which are the site of most intensive evaporation and moreover are the farthest removed from the roots.

In arid regions of India, available soil moisture is used by the roots of annual and perennial plants from the end of rainy season till it is depleted by early summer. Later a partial or total *status quo* is maintained in soil moisture, mainly in the open with the result that water loss is estimated by sheding or reduction of leaves by plants (Sen, 1972). The rain water is the only source of available moisture in north-west Rajasthan desert. Although the monsoon season starts here from mid-June and extends upto October, the rains are very erratic and scanty. Occurrence of rather long intermission of dry spells between successive rain showers, sometimes ranging from a few days to weeks is not uncommon here. Whatever rain water is retained by the soil, is exploited by the roots of annual and perennial species from the months of June-July to November-December. After that, annual species started disappearing due to moisture scarcity in upper soil layers and their inability to exploit moisture from the deeper soil layers through shallow root system. By this time perennial species also started showing various symptoms of water scarcity which were very much reflected by a remarkable reduction in their transpiring surface. Maximum soil moisture during the present study was recorded in rainy season. Due to depletion of soil moisture in October-November, all annuals nearly dried up or exhibited wilting. All annual halophytic species, such as *S. sesuvioides, T. triquetra* and *Z. simplex* completely disappeared in January with depletion of soi! moisture. The winter showers and pre-monsoon rains, though scanty in quantity, played a significant role in improving the water status of soil and plants.

The values of osmotic potentials differ from month to month and from one season to another. These fluctuations in the values of osmotic potential were found to be associated with the variations in soil moisture content and increase in salinity level in soil. So, salinity also increased the value of osmotic potentials. The plant species of saline area showed a wide range varying in values of osmotic potentials (Waisel & Ovadia, 1972). Similarly in the present study, wide range of the osmotic potential has been recorded from one season to the other for all the selected plant species.

All the halophytic plant species studied here could adjust themselves by changing their osmotic potentials rapidly with a greater range due to the change in osmotic potentials of surrounding soil. This is in accordance with Waisel (1972) who stated that it is probably true that a great majority of the halophytic plants belongs to the adjustable group and that their osmotic adjustment occurs rapidly. Recovery from osmotic stress occurs faster in salt accumulating halophytes than in salt enduring ones (Greenway, 1968). Osmotic adjustment or osmoregulation enables plants to maintain growth as plant water potential decreases (Stewart & Lee, 1974; Sen & Mohammed, 1992). Adjustment occurs through decrease in osmotic potential by solute accumulation as leaf water potential decreses, with the net result that the cell turgor dependent processess such as leaf growth and stomatal opening are affected (Sambo & Aston, 1985). Further, it emerges from the present study that during the rainy season, higher moisture in soil and leaching down of the salts by rains resulted in the decreased osmotic potential of the plants. But with the increase in salt concentration and decrease in soil moisture, plants try to adjust themselves to drought by the accumulation of salts. Thus the accumulation of salts in plants increased their osmotic potentials upto the level of highest stress, so that, plants were able to take maximum water during the hot summer, which resulted in the gradual decrease in their osmotic potentials. Some plants exhibited the highest values of osmotic potentials during winter periods associated with low temperature or/and high salinity level.

The leaf water content, which varied considerably in arid zone plants, and in different months, is not necessarily equal due to physiological changes. Ashby & Beadle (1957) observed that dry matter production and moisture content in leaves of *Atriplex nummularia* increased with increasing salinity of the growing media.In the present study leaves of succulent halophytes, such as *S. fruticosa, S. baryosma, S. sesuvioides, T. triquetra* and *Z.simplex* contain more moisture as compared to non-succulents like *C. cretica*. The increase in leaf moisture in succulents can be explained from the increased salinity alongwith water level in the cell sap of such plants, which makes them fleshy, while there in no such mechanism in non-succulent plants. The values of leaf water content are directly associated with soil moisture. When soil moisture increases the values of LWC also increase and *vice versa.* In the present investigation, percentage of leaf water content was highest in rainy season in all the selected species when soil moisture was also high, and lowest during summer season, except in *S. simplex, S.baryosma, T. triquetra, S. sesuvioides* and *Z. simplex*,.which showed lower values during post-monsoon period. This may be due to ephemeral nature of the three latter species, as they nearly completed their life cycle at that time and started drying, while lower value in *S. baryosma* was due to partial sheding of leaves at low temperature.

The values of relative water content increased in rainy season and decreased in summer season in some desert plants as observed by Sen (1973). In the present study also, the highest values of RWC were observed during rainy season and lowest in dry period in all species at both sites and it seems that it is directly proportional to the leaf water content.

Bound water is formed as unfree or combined water. Vaadia *et al.* (1961) had correlated plant hardiness and especially drought hardiness to the relative amount of bound water in the tissue: the more the bound water, the hardier the plant. In the present investigation, maximum bound water was found in *S. fruticosa* followed by *S. baryosma, Z. simplex,* etc. The higher value of BW in *S. fruticosa* and *S. baryosma* proved their hardiness under higher salinity, because they can withstand extreme saline conditions. All the species studied here exhibited higher bound under water and saline stress.

It is clear from the present investigation that the water relation of plants is directly related to the amount of moisture present in the soil.

ACKNOWLEDGEMENTS

Financial assistance received from University Grants Commission in the form of Departmental Special Assistance Programme is gratefully acknowledgement.

REFERENCES

Ashby, W.C. and N.C.W. Beadle (1957): *Ecology* **38**:344-352.

Greenway, H. (1968): *Israel Journal of Botany* **17**:168-177.

Gomez, K.A. and A.A. Gomez (1984): *Statistical Procedures for Agricultural Research*, John Wiley & Sons, New York, USA.

Janardhan, K.V., Murthy, A.S.P., Giriraj, K. and S. Panchaksharai (1975): *Current Science* **44**:390-391.

Jarvis, P.G. (1967): *Annals of Arid Zone* **6**:74--91.

Laarcher, W. (1983): *Physiological Plant Ecology*, Springer-Verlag, Berlin.

Levitt, J. (1956): *Encycl. Plant Physiology*, **3**:650-651.

Oppenheimer, H.R. (1951): *Journal of Ecology*, **39**:358-362.

Piper, C.S. (1950): *Soil and Plant Analysis,* Interscience Publishers, Inc., New York.

Sambo, E.Y. and M.J. Ashton (1985): *Australian Journal of Plant Physiology* **12**:481-486.

Schonbeck, M.V. and T.A. Norton (1979): *Botanical Marina*, **22**:133-144.

Sen, D.N. (1972): In *Ecophysiological Foundation of Ecosystem Productivity in Arid Zone.* International Symposium USSR, PP. 79-83.

Sen, D.N. (1973): *Vegetatio* **27**:201-265.

Sen, D.N. (1982): *Plant Life in Indian Desert.* Geobios International, Jodhpur, India.

Sen, D.N. and S. Mohammed (1992): In *Role of Biotechnology in Agriculture* (Prasad, B.N. : Ghimire, G.P.S. and V.P. Agarwal, eds.) Oxford & IBH Publishing Co. Pvt. Ltd., New Delhi, pp. 129-137.

Sen, D.N. and S. Mohammed (1994): In *Handbook of Plant and Crop Stress* (M. Pessarakli, ed.) Mercel Dekker, Inc., New York, pp. 125-145.

Sen, D.N., Chawan, D.D. and K.D. Sharma (1972): *Flora* **161**:463-471.

Stayer, R.O (1960): *Bulletin of Research Council Israel*, **8**:159-168.

Stewart, G.R. and J.A. Lee (1974): *Planta* **120**:279-289.

Vaadia, Y., Raney, F.C. and R.M. Hegen (1961): *Annual Review of plant Physiology*, **12**:265-293.

Waisel, Y. (1972): *Biology of Halophytes.* Academic Press, New York.

Waisel, Y. and S. Ovadia (1972): *Israel Journal of Botany* **21**:42-52.

Weatherley, P.E. (1950): *New Phytology*, **49**:81-87.

Experimental Studies on the use of Zinc Tailings as Inorganic Fertilizers

Y.D. Tiagi
Department of Botany,
ML Sulhadia University, Udaipur.
&
Shashi Kshetrapal
Department of Botany,
University of Rajasthan, Jaipur.

INTRODUCTION

The production of large amount of wastes, the tailing, is an obvious feature of Ore-beneficiation in most mineral processing industries. The tailing are usually dumped in depressions called as tailing dams. At Zawar Mines, though definite evidence is available for a zinc smelting industry as long ago as 170 ± 60 B.C. (Craddock *et al.*, 1983) large scale mining along modern lines started only a few decades ago. Here, out of the 13 tons of zinc ore deposited in low lying area. The old tailing dam having a surface area of 3 8.5 ha is already full to the brim. The tailings contain very finely ground material and toxic quantities of metals like zinc, lead and cadmium (Table 1). During the dry season, the finer particles are blown over vast areas leading to heavy metal pollution of land, water, food as well as fodder. The magnitude of this pollution can be gauged from the fact that the Mangula dams in Rhodesia are stated to lose 86 tones per day as dust (Down & Stocks, 1977). Because the tailing contain some of the metals which are essential for plants it appeared promising to carry out experimental studies in order to find out the feasibility of their utilization as inorganic fertilizers.

MATERIAL AND METHOD

Three crop plants, namely *Glycin max.* (L.) Merr. (White seeds), *Hordeum vulgare* L. and *Lathyrus sativus* L. were grown in pots in the present experimental studies. Seeds of the former two species were, obtained from the seed corporation of India, Udaipur and those of *Lathyrus sativus* from the local market in Sagar (M.P.), through the courtesy of Dr. T.R. Sahu. Three seeds of each crop were grown in each pot filled with field soil containing 00,2.5,5,7.5 and 10% zinc tailings. The experiments was laid in Latin square design with three replicates for each treatment. A separate set of three replicates of each treatment was also kept for counting the number of seeds per plant after it reached full maturity. Three healthy seeds were inserted equidistantly in each pot. Pots were sparingly irrigated daily up to 10 days after germination and then at intervals of 48 hours in *Glycine* and *Hordeum* but 72 hours in *Lathyrus.* The crop was harvested after a period of 90 days in *Glycine* and *Hordeum* and 75 days in *Lathyrus* for the determination of dry weight of shoot after drying it for 72 hours at 80^{0}C. Total chlorophyll and nitrogen content of leaves were determined, respectively by the method of Vernon (1960) and the microkjeldahl's method (Misra, 1968). Elemental analysis was done on atomic absorption spectrophotometer. The data were subjected to relevant statistical analysis.

OBSERVATIONS

Data on the physical and chemical analysis of zinc tailings and the experimental soil are given in table 1 and on the effect of tailings on shoot biomass (dry matter), chlorophyll and nitrogen content of leaves at the flowering stage of the shoot in tables, 2-5. In a separate study, it was found that the uptake of Zn, Cd and Pb in the shoot of the experimental plants was only negligibly higher than in the controls.

Table 1: Physico-chemical characteristics of the experimental soil and Zawar Zinc tailings.

Sl.No.	Characteristics	Experimental soil	Zawar Zinc Tailing
1.	pH	7.1-7.5	7.3-7.8
2.	Clay (%)	2 ±001	nil
3.	Silt (%)	4 ±006	nil
4.	Fine sand (%)	85 ±5.550	43.4±3.311
5.	Coarse sand (%)	9 ±4.550	56.7±3.311
6.	Grit (%)	nil	nil
7.	Specific gravity (%)	1.837	2.740
8.	Water holding capacity (%)	15.396	13.883
9.	Porosity (%)	119.823	35.621
10.	Organic carbon	00.872±00.213	nil
11.	Ammonium acetate exchangeable:		
	Zinc (ug/g)	13.330±2.113	63.50±2.19
	Cadmium (ug/g)	ND	7.500±1.323
	Copper (ug/g)	2.113±0.022	18.452±1.185
12.	HCL exchangeable:		
	Zinc ug/g	25.892±3.982	1790.670±21.335
	Cadmium ug/g	ND	9.992±1.327
	Copper ug/g	5.298±2.081	27.392±3.899
13.	Water soluble:		
	Zinc ug/g	0.257±0.001	2.110±0.002
	Cadmium ug/g	ND	1.260±0.002
	Copper ug/g	0.005±0.001	0.500±0.001

ND = Not Detectable.

Table 2: The efficacy of different zinc tailing loads on shoot biomass, total chlorophyll and nitrogen content in 75 days old plants of *Glycine max*.

Sl.No.	Treatment Percentage tailing load	Shoot Biomass (g^{dm})	Total Chlorophyll (mg/g)	Total nitrogen (%)
1.Control	0.00	6.81 ± 0.41	5.16 ± 0.06	2.06 ± 0.05
2.	2.5	8.18 ± 0.13	6.07 ± 0.06	2.94 ± 0.07
3.	5.0	10.64 ± 0.29	7.07 ± 0.12	4.08 ± 0.15
4.	7.5	9.21 ± 0.19	5.09 ± 0.01	3.01 ± 0.02
5.	10.0	7.01 ± 0.77	3.87 ± 0.12	1.10 ± 0.02
SEM		0.25	0.05	0.04
CD				
5%		0.7877	0.1575	0.1260
1%		1.1204	0.2240	0.1792

ANOVA TABLES:

(a) Shoot Biomass

		DF	SS	MSS	F
i)	Within the treatment	4	30.4712	7.62	42.26**
ii)	Among the treatment	10	1.8028	0.18	

(b) Total Chlorophyll

		DF	SS	MSS	F
i)	Within the treatment	4	17.2081	4.30	605.86**
ii)	Among the treatments	10	0.0710	0.01	

(c) Total Nitrogen

		DF	SS	MSS	F
i)	Within the treatment	4	14.9265	3.73	625.96**
ii)	Among the treatments	10	0.0710	0.01	

Correlations:	r-value
Biomass: Chlorophyll	0.746**
Biomass: Nitrogen	0.890**

Table 3: The efficacy of different loads of zinc tailings on dry matter production, total chlorophyll and nitrogen content in 75 days old plants of *Hordeum vulgare*

Sl.No.	Treatment Percentage tailing load	Shoot Biomass (g^{dm})	Total Chlorophyll (mg/g)	Total Nitrogen (%)
1.Control	00.0	98.53±0.63	4.63±0.06	1.76±0.04
2.	2.5	100.40±0.02	5.00±0.01	2.30±0.03
3.	5.0	103.20±0.01	4.04±0.13	3.08±0.16
4.	7.5	99.94±0.07	3.03±0.06	2.57±0.05
5.	10.0	91.16±0.17	2.07±0.06	1.93±0.06
SEM		0.17	0.04	0.05
CD				
5%		4.222	0.126	0.157
1%		6.005	0.179	0.224

ANOVA TABLES:

(a) Shoot Biomass

		DF	SS	MSS	F
i)	Within the treatment	4	244.4337	61.11	706.27**
ii)	Among the treatment	10	0.8652	0.09	

(b) Total Chlorophyll

		DF	SS	MSS	F
i)	Within the treatment	4	17.2837	4.32	828.78**
ii)	Among the treatments	10	0.0521	0.01	

(c) Total Nitrogen

		DF	SS	MSS	F
i)	Within the treatment	4	3.3234	0.83	117.60**
ii)	Among the treatments	10	0.0707	0.01	

Correlations:	r-value
Biomass: Chlorophyll	0.692**
Biomass: Nitrogen	0.703**

Table 4: The efficacy of different loads of zinc tailings on dry matter productio total chlorophyll and nitrogen content in 50 days old plants of *Lathyrus sativus*

Sl.No.	Treatment Percentage tailing load	Shoot Biomass (g^{dm})	Total Chlorophyll (mg/g)	Total nitrogen (%)
1. Control	00.0	0.89±0.08	4.28±0.02	19.08±0.02
2.	2.5	1.00±0.01	4.56±0.04	19.97±0.12
3.	5.0	0.80±0.01	4.23±0.05	19.00±0.01
4.	7.5	0.61±0.02	3.27±0.04	18.05±0.06
5.	10.0	0.40±0.02	3.08±0.14	15.07±0.04
SEM		0.02	0.04	0.04
CD				
5%		0.0630	0.1260	1.2918
1%		0.0896	0.1792	1.8374

ANOVA TABLES:

(a) Shoot Biomass

		DF	SS	MSS	F
i)	Within the treatment	4	0.6819	0.17	129.87**
ii)	Among the treatments	10	0.0131	0.00	

(b) Total Chlorophyll

		DF	SS	MSS	F
i)	Within the treatment	4	5.2814	1.32	248.59**
ii)	Among the treatments	10	0,0531	0.01	

(c) Total Nitrogen

		DF	SS	MSS	F
i)	Within the treatment	4	43.0416	10.76	2591.57**
ii)	Among the treatments	10	0.0415	0.00	

Correlations:	r-value
Biomass: Chlorophyll	0.951**
Biomass: Nitrogen	0.946**

Table 5: The efficacy of different loads of zinc tailings on seed number per plant and weight (g) of 100 seeds.

Sl.No.		Treatment Percentage tailing's load	Plant Species: *Glycine max.* Weight/ 100 seeds	*Glycine max.* Seed number/ plant	*Lathyrus sativus* Weight/ 100 seeds	*Lathyrus sativus* Seed number/ plant	*Hordeum vulgare* Weight/ 100 seeds	*Hordeum vulgare* Seed number/ plant
1.	Control	0.01	18.36±1.07	44.33±2.52	10.04±0.13	34.33±2.52	6.10±0.01	26.00±1.00
2.		2.50	20.16±0.64	50.00±2.65	10.01±0.17	38.00±1.00	6.16±0.06	30.33±1.16
3.		5.00	19.73±0.55	66.67±3.21	11.05±0.10	41.00±1.00	6.51±0.08	36.33±0.58
4.		7.50	17.33±0.58	41.00±2.00	10.07±0.17	38.00±1.00	6.88±0.06	40.00±1.00
5.		10.00	14.35±0.56	16.67±1.53	9.98±0.13	31.67±1.15	7.03±0.06	41.67±0.58
	SEM		0.41	1.41	0.08	0.84	0.03	0.52
	CD							
	5% level		1.2918	1.4427	0.2520	2.6467	0.0945	1.6384
	1% level		1.8374	6.3191	0.3585	3.7645	0.1344	2.3304

ANOVA TABLES

	DF	SS	MSS	F
(a) *Glycine max*				
Seed Weight:				
i) Within the treatment	4	64.5094	16.13	32.43
ii) Among the treatments	10	4.9723	0.50	
Seed number/plant				
i) Within the treatment	4	3916.9333	979.23	163.21**
ii) Among the treatments	10	60.0000		
(b) *Lathyrus sativus*				
Seed Weight:				
i) Within the treatment	4	2.5414	0.64	31.52**
ii) Among the treaments	10	0.2016	0.02	
Seed number/plant				
i) Within the treatment	4	158.2667	39.57	18.55**
ii) Among the treatments	10	21.3333	2.13	
(c) *Hordeum vulgare*				
Seed Weight:				
i) Within the treatment	4	2.1193	0.53	155.17**
ii) Among the treatments	10	0.0341	0.00	
Seed number/plant				
i) Within the treatment	4	521.7333	130.43	163.04
ii) Among the treatments	10	8.0000	0.80	

Correlations:	-value		value
Seed weight		Seed number/plant	
Glycine: Lathyrus	0.376 NS	Glycine: Lathyrus	0.845**
Glycine: Hordeum	0.751**	Glycine: Hordeum	0.405 NS

DISCUSSION

Maze (1919), For the first time, brought forward convincing evidence to show the essentiality of zinc for plant growth. However, it was not fully accepted until the early thirties when Chandler *et al.* (1931, 1932) working with peaches, and Johnston (1933) and Parker (1934) working with *Citrus* were able to correct little leaf of peaches and mottle leaf of *Citrus*, respectively by the application of zinc compounds. Hoagland *et al.* (1935, 1936) and Chapman *et al.* (1937) could produce the symptoms of zinc deficiency under culture conditions and then correct the application of zinc compounds. The symptoms of zinc deficiency are varied and may involve a combination of chlorosis, rosetting, dieback and depressed or abnormal vegetative growth. Deficiency symptoms are most severe in fruit crops; beans, soyabean, corn, hops, lim a beans, flax and castor beans among others have been found to be most sensitive to zinc deficiency (Viets, 1966).

In his recent review of the essential and functional metals in plants, Hewitt (1983) has pointed out that zinc is an absolutely essential elements for plant growth and that zinc is essential for the activity of more than 20 enzymes. Price *et al.* (1972) have thoroughly reviewed the role of zinc in plant life. Carbonic anhydrase which catalyses the hydration of carbon dioxide contains zinc (Edwards and Hassall, 1980). The essential role of zinc in metabolism has also been discussed by various other authors (Bowen, 1966; Harrison and Hoare, 1980). Several other workers have shown the association of zinc with cell proteins (Keilin and Mann, 1939; Hoch and Vallee, 1957; Skoog, 1940; Tsui, 1948; Wildman *et al.,* 1947;, Nason, 1950; Shervin & Purves, 1969). Thus, the essentiality of zinc in plant life is a proved fact. It is due to this fact that certain concentration of zinc when added to soils lead to increased crop productivity (Bano and Tiagi, 1991; Sarkar *et al.,* 1990; Ganguli *et al.,*1981).

According to Brown and Krantz (1966) and Grewal *et al.,* (1969) soil containing 0.69 ppm or less of zinc may be considered as zinc deficient. On the contrary, certain plants species as a result of evolutionary modification to metalliferous soils have developed the capacity to absorb and accumulate unusually high quantities of zinc and other metals (Tiagi, 1990; Tiagi and Aery, 1981, 1982). However, for most crops and plants a level of 100-200 ppm of soil zinc is suitable for optimal growth.

Zawar zinc tailings contain appreciable quantities of zinc and a few other metals. Present experimental studies have shown that these tailings can be used to supply the zinc needs of crop plants and can therefore be used as inorganic additives to check the appearance of zinc deficiency symptoms. As can be inferred from data in Tables 2-5, there is a significant increase in the amount of shoot biomass, total leaf chlorophyll and nitrogen content at the time of initiations of flowering. The treatments, statistically are very highly significant. The loading of soil with zinc tailings in most of the treatments at the rate of 5% and in a few cases at the rate of 2.5% has given the most superior results. Thus, these tailings could be packed in bags and used as zinc micronutrient fertilizers in all zinc-deficient soils. A pollutant at one place could, thus be utilized as a fertilizer at another place.

ACKNOWLEDGEMENTS

The senior author is grateful to the University Grants Commission, New Delhi for financial assistance. We are also grateful to Dr. Ms. Sandhya Tyagi of Sukhadia University, Udaipur for taking care of the experimental pots in the Botanical Garden at Udaipur.

REFERENCES

Bano, F.and Y.D. Tiagi (1991): *Indian Bot. Cont.* **8**:1-7.

Bowen, H.G.M. (1966): *Trace elements in Biochemistry,* Acad. Press, London.

Brown, A.L. and B.A. Krantz (1966): *Proc. Amer. Soil. Sci.* **30**:86-89.

Chandler, W.H., Hoagala, D.R. and P.L. Hibbard, (1931): *Proc. Amer. Soc. Hort. Sci.* **28**:556-560.

Chandler, W.H., Hoaland, D.R. and P.L. Hibbard (1931): *Proc. Amer. Soc. Hort. Sci.* **29**:259-263.

Chapman, H.D., Vanselow, A.P. and G.F. Jr. Liebig (1937): *Agr. Res.* **55**:365-379.

Craddock, P.T., Gurjar, L.K. and K.T.M. Hegde (1983): *World Archeol.* **15**:211-213.

Craddock, P.T., Freestone; I.C., Gurjar, L.K., Hegde K.T.M. and V.H. Sonawane (1985): *Mining Magazine* **145**:45-52.

Down, C.G. and G. Stocks (1977): *Mining Magazine,* **117**:25-33.

Edard, N.A. and K.A. Hassall (1980): *Biochemistry and Physiology of the cell.* Mcgraw-Hill Book Company London.

Grewal, I.S., Randhawa, N.S. and D.R. Bhumbla (1969): *J. Res. Punjab Agr. Uni.* **36**:38-45.

Harrison, P.M. and R.J. Horse (1980): *Metals in Biochemistry* Chapman and Hall, London, NY.

Hewitt, E.J. (1983): In: *Metals and Micronutrients, Uptake and Utilli-zation by Plants* Robb. D.A. and P rpoint, W.S. (Eds.) Acad. Press NY, pp. 277-323.

Hoagland, D.R., Chandler, W.H. and P.L. Hibbard (1935): *Proc. Amer. Soc. Hort. Sci.* **33**:131-141.

Hoagland, D.R.; Chandler, W.H. and P.R. Stour (1936): *Proc. Amer. Soc. Hort. Sci.* **34**:210-212.

Hoch, F.L. and B.L. Valle (1957): In Lamb, C.A. Bentley O.G. and Beattic, I.M. (eds.) *"Trace Elements."* Acad Press. NY.pp. 337-363.

Johnston, I.C. (1933): *Citrograph* **18**:116-118.

Kelin, D. and T. Mann (1939): *Nature* **144**:442-443.

Maze, P. (1919): *Ann. Inst. Pasteur* **33**:139-173.

Misra, R. (1968): *Ecology workbook* pp. 244, Oxford and BH publishing Co., Calcuttta.

Nason, A. (1750): *Science* **112**:111-112.

Parkaer, E.R. (1934): *Proc. Amer. Soc. Hort. Sci.* **31**:98-107.

*Price, C.A., Clark, H.E. and E.A. Funkhouser (1972): In *Micronutrients in Agriculture.* "Mortvedt, I.I. Giordano P.M. and Lindsay, W.I. (Eds.) *Soil. Sci. of America, Madison.* pp.231-242.

Shervin, I.E., and W.K. Purves (1969): *Plant Physiol.* **44**:1303-1309.

Skoog, F. (1940): *Amer. J. Bot.* 27:239-591.

Tiagi, Y.D. (1990): *Geobotany and biogeochemistry in mineral prospecting.* Presidential address Bot. Sec. 77th Sess., ISCA, Calcutta.

Tiagi, Y.D. and N.C. Aery (1981): *Proc. Indian Nat. Sci. Acad.* **47B:** 867-887.

Tiagi, Y.D. and N.C. Aery (1982): *Vegetation*. **50:**65-70.

Tobin, A.J. (1969): In *"CO_2 Chemical biochemical and physiological aspects",* Forester, R.E. I.T. Otis A.B. and Roughton, F.J. (eds.) NASA Washington. p.139.

Tsui, C. (1948): *American J. Bot.* **35:**309-311.

Vernon, L.P. (1960): *Anal. Spring field*, **32:**1144-1150.

*Viets, F.G. Jr. (1966): In *Zinc Metabolism*, Prasad, A.S. (ed.) Thomas Spring field, pp. 90-128.

Viets, F.G. (Jr.), Boawn, L.C. and C.L. Crawford (1954): *Plant physiol.* **29:**76-79.

Wildman, S.G., Ferri, M.G. and I. Bonner (1947): *Arch. Biochem.* **13:**131-144.

---

* *Not seen in original.*

Biological Effects of Fusarial Toxins

K.S. Bilgrami and K.K. Sinha
University Department of Botany,
Bhagalpur University,
Bhagalpur -812 007.

Fusaria are widespread in nature, occurring as facultative saprophytes/parasites on a variety of plants and plant parts. Many of the Fusarial species can elaborate toxins of varying chemical compositions and dimensions which are generally found as natural contaminants in various types of animal foods and feeds. When ingested, these *Micotoxins* can lead to severe disorders in animals as well as the human beings. These also cause numerous plant diseases including wilting.

Involvement of Fusarial toxins in plant disorders was first highlighted in 1945 by Plattner and Clauson Kass. They observed the wilting of tomato due to lycomarasmin produced by *Fusarium lycopersici.* Gaumann *et al.* (1952) isolated another toxin from the culture filtrates of *F. lycopersici* and named it as fusaric acid which is considered to play a vital role in plant pathogenesis and causing wilt of tomato and cotton and *bakane* disease of rice. A number of diseases in wheat have also been reported from Eastern Europe and different parts of Russia. These include the seedling blight and root rots of plants (Kaczalowa and Jacenko, 1974; Kaczalowa and Malarenko, 1974; Srobar, 1978).

The role of Fusarial toxins in the ailments of animals and human beings came to light due to widespread occurrence of *Alimentary Toxic Aleukia* (ATA) in Western Siberia during the world war II. The disease was attributed to be caused due to the toxins of trichothecene group which were elaborated by *F. sporotrichiodes* and *F. poae*. Symptoms of disease included severe haemorrhage, leukopenia, necrotic angina and degeneration of bone marrows. The sufferings of similar nature were earlier reported in 1913 and 1932 in many parts of Eastern Siberia where entire families perished and populations of some villages were decimated due to the consumption of over wintered grains heavily infested with the above mentioned species of *Fusarium.* Since that time, the disease sprang up regularly in the agricultural regions of Urals, Ukaraine and Central Asia.

Mouldy corn *toxicosis* is the most important Fusarial toxin-induced disease in animals. This was reported for the first time in the United States where more than 20% pigs died after consuming *Fusarium* contaminated feed (Sippel *et al.,* 1953). The toxin associated with such epidemics were deoxynivalenol and nivalenol. Refusal and aversion to food, digestive disorders and diarrhoea were the common symptoms. The significance of mouldy corn toxicosis in human health seems to be rather limited. In Japan, there were cases of nausea, vomiting and diarrhoea associated with the consumption of *Fusarium* contaminated wheat and rice but these have frequently been referred to as redmould disease involving *F. graminearum.*

Toxigenic Fusaria

About 200 strains, belongings to 20 different species of *Fusarium*, are known to be toxigenic (Marasas *et al.,* 1984). These have been isolated from different consumable items like fruits, vegetables and cereals. The major toxins elaborated by them include zearalenone and the trichothecenes. Some other metabolites are also produced by them. The common toxigenic Fusaria and the types of their toxins are as follows:

F. sporotrichioides - The mycotoxins produced by the isolates include butenolide and several trichothecenes like diacetoxyscirpenol, T-2 toxin, deoxynivalenol (DON) and nivalenol.

F. poae - Toxigenic strains of *F. poae* have also been frequently referred to as *F. tricinctum.* The mycotoxins elaborated are butenolide, diacetoxyscripenol and T-2 toxins.

F. equiseti - This species is also known as *F. roseum.* It produces butenolide and eight trichothecenes including diacetoxyscripenol, nivalenol and T-2 and zearalenone.

F. graminearum - Some trichothecenes (DON, nivalenol etc.) and zearalenone have been reported to be produced by the toxigenic strains.

F. moniliforme - Mycotoxins produced by this species include fusaric acid, fusarins, fusariocins, moniliformin, fumonisins and zearalenone.

F. acuminatum - It is known to elaborate T-2, Ht-2, diacetoxyscirpenol, neosolaniol etc.

F. nivlae - Elaborate nivalenol and fusarenone-X.

F. lycopersici - Capable of producing lycomarasmin and fusaric acid.

Natural Occurrence of Fusarial Toxins

The common Fusarial toxins encountered as natural contaminants of food and feed substrates include zearalenone (ZEA), deoxynivalenol (DON), nivalenol and T-2 toxin.

A very high incidence of ZEA has been reported in the maize samples from Austria (Lew *et al.,* 1979) and France (Collet and Regnier, 1977) and in visibly damaged wheats from Germany (Dietrich *et al.,* 1988). A broad survery on the incidence of ZEA from several European countries revealed a varying percentage of contaminated cereal grains like maize (26%), wheat (25%), barley (19%), oats (15%) and mixed feed (12%). Few recent reports from India also indicated the occurrence of ZEA in freshly harvested maize (Sinha, 1991) and sorghum and ragi (Ansari and Shrivastava, 1991).

Natural occurrence of trichothecene has also been reported from widespread areas. Morooka *et al.* (1972) isolated DON and nivalenol from barley and wheat in the Kagawaken region of Japan. A very high incidence of DON has been reported in wheat from Germany (Noonpugde, 1985). Maize been frequently found to contain DON in substantially high amounts (Yoshizawa and Morooka, 1977; Vesonder *et al.,* 1978; Jemmali *et al.,* 1978). Fragmentary reports from India also reveal the occurrence of trichothecene in cereals. Bhat *et al.* (1989) found the samples of wheat and wheat products which were the cause of fusariotoxicosis in humans in Kashmir valley. Bhavanishankar and Shantha (1987) reported T-2 in peanut, sorgum and maize samples from Mysore region. From Bihar, Ansari and Shrivastava (1991) observed DON and T-2 toxin in sorghum, ranging from 12 to 53 μg/kg and 9.0 to 16 μg/kg, respectively.

Biological Effects of Fusarial Toxins

(A) **Animal System**

Most of the naturally occurring fusarial toxins have been reported to affect the general physiology of laboratory animals. We shall, however, confine our discussion on the effect of three important toxins i.e., zearalenone, deoxynivalenol and T-2 on animal system.

Effect of Zearalenone (ZEA)

ZEA is characterised by its most important oestrogenic properties which include swelling of vulva, prolapse of vagina, metrorrhagia, increase in secretion and size of uterus, overgrowth and lactation in mammary glands. Dietary ZEA acts systematically producing an "estrus syndrome" in swine, cattle and other laboratory animals (Chang *et al.,* 1979). The clinical effects of subacute ZEA mycotoxicosis can be decreased feed intake, reduced growth and decreased reproduction. ZEA has also been implicated in teratogenesis and carcinogenesis (Microcha *et al.,* 1980).

Of the domestic animals, swine is the most sensitive and susceptible to ZEA feeding. In an experiment with young pigs, Morocha and Christensen (1974) reported that a dose of only 0.02 mg/kg. b.w. caused vulvar swelling and high dose resulted in a variety of signs of hyperestrogenism. Mirocha *et al.,* (1977) have also reported ZEA toxicity in livestock and other experimental animals when mould contaminated feed was ingestd by them. ZEA was observed to cause sows to farrow smaller litters and smaller pigs when added to the diet of pregnant sows at 25 and 50 mg/kg of feed (Chang *et al.,* 1979). ZEA toxicosis and metabolism have also been investigated in rat (Smith, 1990, 1982; James and Smith, 1982).

With bull, Vanyi *et al.* (1980) found that continuous feeding of corn containing 20 ppm ZEA for 21 days had made the semen of bulls unfit for insemination and this could not be reversed by even 4 months feedings on ZEA free diet. Histopathological observations showed degeneration of germinal epithelium and a 75% degeneration of sperms.

Poultry, especially chickens, are comparatively less susceptible to ZEA. Young chickens when given high doses of ZEA for 7 days showed no gross effect at 50 mg/kg b.w. (Sherwood and Peberdy, 1976). Marks and Becon (1976) reported that even the supply of ZEA for 8 weeks to chicken at dietary levels upto 30 ppm (.2 mg/kg b.w.) had no effect on body weight gain. Levels upto 800 ppm in the diet (approx.59 mg/kg.b.w.) also did not show any effect on egg laying or the size of eggs whereas very high dietary levels (100 ppm) caused a 20% reduction in egg production in turkeys (Allen, 1980; Allen *et al.,* 1981).

Effects of T-2 toxin

T-2 toxin is one of the most toxic of all the known thrichothecenes. This toxi is important due to the outbreak of epidemic of mycotoxicosis in human beings and a large number of animals. T-2 toxin has been conjectured to be a factor in fatal alimentary toxic aleukia in human beings (Joffe, 1983) and has been implicated in other mycotoxicoses such as red mould disease of human beings and animals as bean hull poisoning in horses in Japan (Uneo *et al.,* 1972), fusariotoxicosis in poultry (Greemway and Puls, 1976) and mouldy corn toxicosis in cattle (Hsu *et al.,* 1972). The application of a small amount of T-2 toxin on the skin can cause servere inflammations and irritations (Wel *et al.,* 1972; Chunh *et al.,* 1974; Eppley *et al.,* 1974; Hayes and Schiefer, 1979). Bamburg *et al.,* (1968) reported the swelling of the treated area within hours of applications of T-2 toxin on to the treated shaved skin of 21 days old white rat followed by a dry, crusty appearance of the skin which had become dark and purple later on. Histological examinations of the inflammed area of rats showed necrosis of the epidermal cells in several other studies (Marasas *et al.,* 1969). The application of large amount of T-2 toxin on the skin has, however, been reported to cause death within 48 hours without any noticeable skin reaction (Bamburg and Strong, 1971).

Besides the irritational effects on skin, T-2 toxin is also capable of causing necrosis of the intestinal mucosa and spleenic follicles when consumed with the diet (Hayes, 1979). The supply of meals containing 12 ppm of this mycotoxin at the time of breeding and gestation intervals have been reported to cause congestion and oedema of the gastrointestinal mucosa and a decrease in reproductive efficiency in sows (Kurtz, 1981). When the toxin was administered intraperitonially, it resulted absorption in the pregnant sows without any histopathological lesions in the placenta or the foetus (Weaver *et al.,* 1987, a,b). This mycotoxin has also been reported to cause congestion and haemorrhage in the stomach, small and large intestine when lethal doses were administered to the swine (Beasly *et al.,* 1988). Histopathological abnormalities have also been recorded due to T-2 toxin in rabbit (Niyo *et al.,* 1988). Centrilobular hepatocellular swelling in the liver was observed when T-2 toxin was ingested orally at 0.25 mg/kg/day for 3 weeks. After the administration of T-2 toxin at 0.75 mg/kg/day for 11 days, portal and periportal fibrosis in the liver and severe lymphoid depletion of thymic cortex were recorded frequently. The domestic cat has also been found to be very susceptible to this mycotoxin. At a dose level of 0.1-0.2 mg/kg, it manifests major gross clinical signs including vomiting, diarrhoea, anorexia, ataxia of hind legs, discharge from the eyes and ejection of hemorrhagic fluid (Sato *et al.,* 1975, Cole and Cox, 1981). In calves, Pier *et al.* (1976) reported development of ulcer and an acute enteric response with bloody faces due to oral feeding of 0.32-0.46 mg/kg T-2 mycotoxin. The calves became inappetent, dehydrated and a slight to severe weight loss was observed respective to the dose levels of toxin. T-2 toxin-induced feed refusal has also been observed in pigeon, where it caused vomiting within 10 minutes of exposure (Rukimini *et al.,* 1980; Fairhurst *et al.,* 1987). Richard *et al.* (1978) have reported that turkeys were more susceptible to dietary T-2 toxin than the chickens which developed oral lesions when the toxin was fed for a prolonged period.

Effect of Deoxvnivalenol (DON)

DON is one amongst the less toxic trichothecene group of mycotoxins. But its high frequency in toxic concentration in almost all the food/feed commodities poses a serious threat to consumers. The major effect of DON is on the feed consumption and body weight of the animals. Emesis and diarrhoea have been frequently reported in several studies (Friend, 1982; Yarom *et al.,* 1983; Chavez, 1984). Khera *et al.* (1982) have observed the embryotoxic and teratogenic effects of DON when adminstered to pregnant mice on 8-11 days of pregnancy. Histopathological abnormalities in the thymus of mouse were also recorded when treated daily with 7.5 mg/kg b.w. of DON, where central medullary area and marked cortical atrophy with severely depleted thymocytes were observed (Arnold *et al.,* 1986). In poultry, Huff *et al.* (1981) have observed that the birds began to gasp, became lethargic and dropped their wings from the normal upright position immediately after administration of DON. Subsequently a loss of balance and righting reflex ensued. Acute DON toxicity in the poultry has also been characterised by disturbances of the nervous system similar to those seen during T-2 toxicoses (Wyatt *et al.,* 1973).

Adverse effects of DON feeding have also been observed in rainbow trout (*Salmo gairdneri*) where feed refusal and loss in body weight gains were observed when 20 μg/g or more DON concentration was maintained in the diet for 8 weeks. Hintikka (1989) has also reported histopathological abnormalities in the same trout when commercially pelleted feed containing 1-60 μg/kg DON was provided to them. The abnormalities included perforation of the body cavity, severe haemorrahagic oedema of the gastric and intestinal wall and complete destruction of pancreatic tissue.

Under CSIR project,we also recorded the effect of above three Fusarial toxins on Albino mice (*Mus musculus*). We divided the animals of either *sex* (22-25gm) into three dietary groups for each toxin. The groups and treatments were as follows:

1. **ZEA group I** -given standard toxin-free diet (Gulmohar diet-Hindustan Lever).

   ZEA group II - 0.1ml ZEA, 0.11 mg/ml per 22 g body weight at a dose level of 0.5 mg/kg.

   ZEA group III - 0.20 ± 0.03 mg/kg toxin mixed with the feed.

2. **DON group I** -given standard diet.

   DON group II - 0.1 ml DON, 0.66 mg/ml per 22g body weight at a dose level of 3 mg/kg.

   DON group III - 0.40 ± 0.05 mg/kg toxin mixed with the feed.

3. **T-2 group I** - given standard diet.

   T-2 group II - 0.1 ml T-2, 0.11 mg/ml per 22 g body weight at a dose level of 0.5 mg/kg.

   T-2 group III - 0.10 ± 0.02 mg/kg toxin mixed with the feed.

Animals of group II were given toxin orally, twice a week for eight weeks along with propylene glycol. The animals received toxin 1hr before the regular meal supplied to them. The dose was adjusted in such a way that the total amount of toxin supplied to each animal did not exceed the maximum single lethal dose (Ueno and Ishii, 1985).

Following abnormalities were recorded in the affected mice.

(a) **Clinical abnormalities**

A remarkable effect on body weight gain (b.w.g.) and in the general behaviour of the experimental animals was evident due to the feeding of Fusarial toxins.

Animals of Group-II getting ZEA suffered 100% mortality and death started after 3 weeks of toxin feeding. In most of the cases death was preceded by marked loss in body weight. The animals of Group-III under this toxin treatment, however, did not have any sharp effect on b.w.g. and all of them survived till the end of the experimental period. Clinical signs due to ZEA feeding included the enlargement of urinary bladder.

Abnormalities preceding death due to most of the above treatments included periods of shivering and difficulty in locomotion and the animals were able to move a very short distance much slower than the normal gait. Immediately after administration of DON and T-2, the mice of Group-II began to gasp and became lethargic. Animals, however, recovered partially after 2-3hrs. The Group-III animals that died due to DON feeding became sluggish after 4 weeks of continuous feeding.

(b) **Histopathological abnormalities**

Histological studies of the treated animals showed marked changes in some of the body organs like skin, kidney, liver, intestine and urinary bladder.

Skin: Epidermal cells of the skin were greatly affected by the T-2 toxin. Formation of papilloma was frequently observed in the animals of both the treated groups (Group-I, fed toxin orally; Group-II, toxin administred by mixing with the feed). Necrosis of epidermis as well as intra-epidermal postule formation was also very common. Sometimes total necrosis of epidermal layer was also recorded which was commonly found in the animals of Group-II only. The other toxins did not show any remarkable change in the skin part.

Kidney: A common manifestation due to T-2 and DON feeding was the presence of fairly large areas of neoplastic elements comprising masses of non-differentiated cells with hyperchromatic nuclei which were mainly present in the glomerular region.

Liver: The study of liver showed portal and periportal fibrosis amongst the animals of T-2 toxin treatment only. Bile duct proliferation in the liver was also observed in this case.

Intestine: Swelling of intestine was common amongst the DON and T-2 treated animals. Necrosis of intestinal mucosa was also frequently observed in both the treatment groups.

Urinary bladder: One of the remarkable signs due to ZEA feeding was the enlargement of urinary bladder which had swelled to more than the double its original size. The histopathological abnormalities in the urinary bladder included the loss of epithelial lining, prominent blood vessels and inflammatory cellular infiltration.

(c) **Chromosomal abnormalities in bone marrow cells**

Animals fed on Fusarial toxins showed a number of abnormalities in their chromosomes. The total frequency of occurrence of abnormal chromosomes were 10.0 ± 1.7 and 16.3 ± 2.1% in the animals of Group-II of DON and T-2 toxin treatments, respectively, whereas it was only 3.3 ± 1.0% in each control group. The animals of this group under ZEA treatment did not survive, hence no record of chromosomal changes could be made. Mice of Group-III showed 17.3 ± 2.2, 6.3 ± 1.4 and 13.3 ± 2.0% abnormal metaphase plates for ZEA, DON and T-2 toxin treatments, respectively. ZEA and T-2 in this case, however, exhibited a significant difference with thair respective control groups.

With regard to the types of abnormalities, both gross and individual types were recorded in each treatment group. The changes of the latter were higher than the former i.e., gross type. The individual type comprised 76% of the total abnormal plates which included breaks, gaps and fragmentation of the chromosomes. The gross type abnormalities showed stickiness, clumping, hypoploidy and pulverisation of chromosomes.

(B) **Plant System**

Only a limited number of Fusarial toxins have been reported to affect the plant systems. Diacetoxyscripenol in lower concentration (Less than 2.5 μg/ml) enhanced the development of roots of cress seedlings and was inhibitory at more than 10 μg/ml (Brian *et al.,* 1961). Complete inhibition of seed germination of *Pisum sativum* has been recorded at 4 μg/ml of T-2 toxin (Burmeister and Hesseltine, 1970). Severe wilting, necrosis and reduction in the average weight and length of the seedlings have been observed in plants when the seedlings were dipped in aqueous solution (About 2.5 μg/ml) of T-2 toxin (Marasas *et al.,* 1971). T-2 toxin also inhibits the alpha-amylase activity in germinated barley seeds. Even the addition of growth promoting factor iike gibberellic acid does not alter the inhibitory effect (Nummi *et al.,* 1975).

Bottalico *et al.* (1980) reported almost complete inhibition of the growth of tomato seedlings at 6 μg/ml deoxynivalenol concentration. Lacobelis and Bottalico (1981) observed the destruction of cell membrane permeability in tomato leaf disc by diacetoxyscripenol. T-2 toxin and several other trichothecenes have been reported to cause electrolyte leakage from tomato leaf discs (Palvlovkin *et al.,* 1986).

Cytotoxic effects of *Fusarium* toxin have also been observed in root tips of *Allium cepa* in which the reduction in the frequency of mitosis as well as chromosomal aberrations were noted (Reiss, 1975).

Few reports also indicated the influence of Fusarial toxins on plant harmonal balance. Brian *et al.* (1961) observed the breaking of dormancy of lateral buds in *Vicia sativa* by diacetoxyscripenol. This toxin has also been reported to kill the apical buds (Andel and Fuchs, 1972).

We have also recorded the influence of different concentration of three Fusarial toxins on the physiological and biochemical processes of maize and mung seeds. A remarkable influence of these toxins on seed germination and seedling growth in both these seeds was observed. Inhibition in germination was above 70% in the crops due to T-2 toxin treatment. DON was the next effective toxin followed by ZEA in inhibiting seed germination. Seedlings growth *i.e.*, root and shoot lengths was also reduced to a significant level by most of the treated groups. The study on biochemical changes revealed depletion of chlorophyll, protein as well as nucleic acid content due to treatment with above toxins. Up to ID_{50} (inhibition dose-50%) the changes in protein and chlorophyll contents were remarkably high. T-2 toxin was most effective in causing depletion both on maize and mung. Prominent cytological abnormalities were also recorded due to these three toxins. A decrease in actively dividing cells of onion root tips was noted which was directly influenced by the concentration of the toxins. Different division phases were also affected but the influence in metaphase count was remarkably high in all the treatments.

REFERENCES

Abramson, D., Clear, R.M. and T.W. Nowicki (1987): *Can. J. Plant Sci,* **67**:611-619.

Allen, N.K. (1980): *Poult. Sci.* **59**:1577.

Allen, N.K., Microcha C.J., Aakus-Allen, S., Bitgood J.J., Weaver G. and F. Bates (1981): *Poult. Sci.* **60**:1165-1174.

Andel, V.O.M. and Fuchs (1972): In *Phytotoxins in plant diseases* (Eds. R.K.S. Wood, A. Ballio and A. Graniti). Academic Press, London, pp. 227-247.

Ansari, A.A. and A.K. Shrivastava (1991): In *Proc. Symp. Mycotoxins, Incidence and Human health*(Ed. K.S. Sinha). The Allied Press, Bhagalpur, pp 139-145.

Arnold, D.L., Mc Gurie, P.F., Nera, E.A., Karpinski, K.F., Bickis, M.G., Zawidka, Z.Z., Fernie, S. and R.F. Vesonder (1986): *Food Chem Toxicol.* **24**:935-941.

Bamburg, J.R. and F.M. Strong (1971): In *Microbial Toxins* (Eds. S. Kadis, A. Ciegler an. Ajl.) Vol. VII, Academic Press, New York, pp 207-289.

Bamburg, J.R., Riggs, N.V. and F.M. Strong (1968): *Tetrahedron* **24**:3329-3336.

Beasly, V.R., Swanson S.P., Corley, R.A., Buck, W.B., Koritz, G.D., and H.R. Burmeister (1986): *Toxicon* **24**:13-23.

Bhat, R.V., Shashidhar, R.B., Ramkrishna, Y. and K.L. Munshi (1989): *Lenset* **1**:35-37.

Bhavanishankar, T.N. and T. Shantha (1987): *J. Sci. Food Agric.* **40**:327-332.

Bottalico, A., Learo, P. and A. Visconti (1980): *Phytopath. Medit.* **19**:196-198.

Brian, A.T., Dawkins, A.W., Grove, J.F., Hemming, H.G., Lowe, D. and G.L. Norris (1961): *J. Exp. Bot.* **12**:1-12.

Burmeister, H.R. and C.W. Hesseltine (1970): *Appl. Microbiol.* **20**:437-440.

Busby, W.F. and G.N. Wogan (1981): In *Micotoxins and N - Nitrosa Compounds: Environmental Risks,* Vol. II (Ed.R.C. Shank), Chapter 3. CRC Press, Boca Raton, Fla. pp. 29-45.

Chang, K., Kurtz, H. and C.J. Microcha (1979): *Amer. J. Vet. Res.* **40**:1260-1267.

Chavez, E.R. (1984): *Can. J. Anim. Sci.* **64**:717.

Chung, C.W., Trucksess, M.W., Giles, A.L. and L. Freidman (1974): *J. Assoc. Off. Anal. Chem.* **57**:1121-1127.

Cole, R.J. and R.H. Cox (Eds.) (1981): *Handbook of Toxic Fungal Metabolites.* Academic Press, New York, P.11850.

Collet, J.C. and J.M. Regnier (1977): *Ann. Nutr. Alim.* **31**:447-457.

Dietrich, R., Biber A., Martibauer, E. and G. Terplan (1988): 10, Nat. Mykotoxin workshop, Kulmbach, F.R.G., 31.05-01.06.

Eppley, R.M., Stoloff, L., Trucksess, M. and M.W. Chung (1974): *J. Assoc. Off. Anal. Chem.* **57**:632-635.

Fairhurst, S., Marrs, T.C., Parker, H.C., Scawin, J.W. and D.W. Swanston (1987): *Toxicology* **43**:31-50.

Friend, S.C.E. (1982): Ph.D. Thesis, University of Saskatchewan.

Gaumann, E., Naef-Roth, S. and H. Kobel (1952): *Phytopath.* **20**:1-38.

Greenway, J.A. and R. Puls (1976): *Can. J. Comp. Med.* **40**:12-15.

Hayes, M.A. (1979): Ph.D. Thesis, University of Saskatchewan.

Hayes, M.A. and H.B. Schiefer (1979): *Food Cosmet. Toxicol.* **17**:611-621.

Hintikak, E.L. (1989): In *Fusarium Mycotoxins, Taxonomy and pathogenicity* (Eds. J. Chelkowski). Elsevier Sci. Pub., Amsterdam, pp. 131-138.

Hsu, I.C., Smalley, E.R., Strong, F.M. and W.E. Ribelin (1972): *Appl. Microbiol.* **24**:684-690.

Huff, W.E., Doerr, J.A., Hamilton, P.B. and R.F. Vesonder (1981): *Poult. Sci.* **60**:1412-1414.

Iacobelis, N.S. and A. Bottalico (1981): *Phytopath Medit.* **20**:129-132.

James, J.L. and T.K. Smith (1982): *J. Animal Sci.* **55**:110-118.

Jemmali, M., Ueno, V., Ishii, K., Frayssinet, C. and M. Etienne (1978): *Experimentia* **34**:1333-1334.

Joffe, A.Z. (1983): In *Handbook of food borne diseases of biological origin* (Ed. M. Rochcigle). CRC Press Inc, Boca Raton, Fia. pp. 353-495.

Kaczalowa, Z.P. and E.A. Makarenko (1974): *Dokl. Mosk. S. Ch. Akademii im, Timiriaziewa* **2**:122-130.

Khera, K.S., Whalen, C., Angres, G., Vesonder, R.F. and T. Kuiper-Goodman (1982): *Bull. Environ. Contam. Toxicol.* **29**:487.

Kurtz, H.J. (1981): In *Current Veterinary Therapy-Food Animal Practice* (Ed. J.L. Howard). W.B. Saunders, Philadelphia, pp. 410-411.

Lew, H., Mullner, E., Hagen, R. and M. Gregor (1979): *Bodenkultur* **30**:309-316.

Marasas, W.F.O., Bamburg, J.R., Smalley, E.G., Strong, F.M., Ragland, W.L. and P.E. Degurse (1969): *Toxicol. Appl. Pharmacol* **15**:471-482.

Marasas, W.D.O., Smalley, E.B., Bamburg, J.R. and F.M. Strong (1971): *Phytopath* **61**:1488.

Marasas, W.F.O., Nelson, P.E. and T.A. Tousson (Eds.) (1984): In *Toxigenic Fusarium Species : Identification and Mycotoxicology.* The Pennsylvania State University Press, University Park.

Marks, H.L. and C.D. Bacon (1976): *Poultry Sci.* **55**:1864-1870.

Mirocha, C.J., Pathre, S.V. and C.M. Christensn (1977): In *Mycotoxins* (Ed. I.F.H. Purchase). Elsevier Science Publications, Amsterdam, pp. 129-148.

Mirocha, C.J., Pathre, S.V. and C.M. Christensen (1980): In *Microtoxins in Human and Animal Health* (Eds. J.V. Rodricks, C.W. Hesseltine and M.A. mehlman). Pathotox Pub. Inc. pp. 345-365.

Mirocha, C.J., Pathre, S.V. and C.M. Christensen (1980): In *Advances in Cereal Chemistry* (Ed. Y. Pomeranz). American Association of Cereal Chemists, St. Paul, Minn. pp 159-225.

Morooka, N., Uratuji, N., Yoshizawa, T. and H. Yamamoto (1972): *Hyg. Soc. Japan.* **13**:368-375.

Niyo, K.A., Richard, J.L., Niyo, Y. and Tiffany (1988): *Amer. J. Vet. Res.* **49**:1766-1773.

Noonpugde, Ch. (1985): *Ernahrung/Nutrition,* **9**:850-854.

Nummi, M., Niku-Paavola, M.L. and T.M. Enari (1975): *Biosc.* **7-8**:266-269.

Pavlovkin, J., Mistrik I., Zajcenko, M.A. and J. Mraz (1986): *Biologia* **41**:681-686.

Pier, A.C., Cysewski, S.J., Richard, J.L. and A.K. Baetz (1976): *Proc. US Animal Health Assoc.* **80**:130-148.

Resis, J. (1975): *Cytologia* **40**:703.

Richard, J.L., Cysewski, S.J. and A.C. Pier (1978): *Amer. J. Vet. Res.* **39**:1674-1679.

Rukmini, C., Prasad, J.S. and K. Rao (1980): *Food Cosmet. Toxicol.* **18**:267-269.

Sato, N., Uneo, Y. and M. Enomoto (1975): *Jap. J. Pharmacol* **25**:263-270.

Sherwood, R. and J.L. Peberdy (1976): *Brit Poult. Sci.* **14**:127-131.

Sinha, K.K. (1991): *Mycotoxin Res.* **7A** (11):178-183.

Sippel, W.L., Burnside, J.E. and M.B. Atwood (1953): In *Proc. 10th Ann. Meeting, Am. Vet. Med. Assn.* Toronto, Canada, pp. 174-181.

Smith, T.K. (1980): *J.Animal Sci.* **50**:278-285.

Smith T.K. (1982): *Can. J. Physiol. Pharmacol* **60**:1444-1449.

Srobar, S. (1978): *Poll Nohospodarstuo.* **24**:514-518.

Ueno, Y. and K. Ishii (1985): In *Trichotecenes and other Mycotoxins* (Ed. J. Lacey). J. Wiley & Sons, Chrichester, pp. 307-316.

Vanyi, A., Timar I. and A. Szeky (1980): *Magy. Allatorv. Lapja* **35**:777-780.

Vesonder, R.F., Ciegler, A., Rogers, R.F., Burbridge, K.A., Bothast, R.J. and A.H. Jensen (1978a): *Appl. Env. Microbiol.* **36**:885-888.

Weaver, G.A., Kurtz, H.J., Bates, F.Y., Chi, M.S., Mirocha, C.J., Behrens, J.C. and T.S. Robinson (1978a): *Vet. Res.* **103**:531-535.

Wever, G.A., Kurtz, H.J., Mirocha, C.J., Bates, F.Y., Behrens, J.C. and T.S. Robinson (1978b): *Can. Vet. J.* **19**:310-314.

Wei, R.D., Smalley, E.B. and F.M. Strong (1972): *Appl. Microbiol.* **23**:1029-1030.

Wyatt, R.D., Colwell, W.M., Hamilton, P.B. and H.R. Burmeister (1973): *Appl. Microbiol.* **26**:757-761.

Yarom, R., More, R., Scherman, Y. and B. Yagen (1983): *British J. Exp. Pathol.* **64**:570-577.

Yoshizawa, T. and N. Morooka (1977): In *Mycotoxins in Human and Animal Health* (Eds. J.V. Roodricks, C.W. Hesseltine and M. A. Mehiman). Pathotox Publishers Inc., Park Forest. pp. 309-321.

INDEX